T0137345

T-Labs Series in Telecommunication Services

More information about this series at http://www.springer.com/series/10013

Abdulbaki Uzun

Semantic Modeling and Enrichment of Mobile and WiFi Network Data

Abdulbaki Uzun
Telekom Innovation Laboratories,
Technische Universität Berlin
Berlin
Germany

ISSN 2192-2810 ISSN 2192-2829 (electronic)
T-Labs Series in Telecommunication Services
ISBN 978-3-030-08095-2 ISBN 978-3-319-90769-7 (eBook)
https://doi.org/10.1007/978-3-319-90769-7

Printed on acid-free paper

This Springer imprint is published by the registered company Springer International Publishing AG part of Springer Nature
The registered company address is: Gewerbestrasse 11, 6330 Cham, Switzerland

Acknowledgements

Working on the doctoral thesis was a tough endeavor, especially after I changed my job in between and the thesis became my “private” matter. However, seeing that there is light at the end of the tunnel after long years of hard work, makes me very emotional and proud.

First of all, I want to thank God, the Most Beneficent and the Most Merciful. Without Him I could not achieve anything. In times of desperation, I knew that there was a door to knock on.

Secondly, I would like to express my gratitude to Prof. Dr. Axel Küpper who gave me the opportunity to work in his team and provided the professional environment to do my research. He always motivated me to pursue a doctor’s degree. In the first five years of my professional career, I learned so much working at the research group *Service-centric Networking* and gained valuable experience for the future.

In addition, my appreciation goes to Prof. Dr. Atilla Elçi and Prof. Dr. Thomas Magedanz for their support and guidance. Moreover, I would like to thank my colleagues at *Service-centric Networking* and the students who collaborated with me and supported my research. Furthermore, I do not want to forget Hans Einsiedler from *Telekom Innovation Laboratories* who was a mentor to me at work and played a major role in my decision to finish this thesis.

I want to say a special and sincere “thank you” to my parents, especially my lovely mother. They always supported and helped me during my entire academic as well as professional career, and they were always there for my little family and me.

The persons who I owe the most debt of gratitude are my wife Berrin and my two lovely children Hubeyb and Meryem. Berrin never left me alone; she always believed in me and supported me since day one of our marriage. Especially the last year, which was very exhausting and not that easy, she never gave up on my professional goals. I love you all so much; all my efforts are just for you three!

Last but not least, I want to thank my parents in law, my grandmother, my other family members, and my friends.

Publications

Here, the author presents a selection of his publications that illustrate the scientific relevance of his contribution within this doctoral thesis.

Book Chapters

[B1] A. Uzun and G. Coskun. Semantische Technologien für Mobilfunkunternehmen - Der Schlüssel zum Erfolg? In *Corporate Semantic Web - Wie semantische Anwendungen in Unternehmen Nutzen stiften*, pages 145–165. Springer, Berlin, Heidelberg, 2015.

Journals

[J1] A. Uzun, E. Neidhardt, and A. Küpper. OpenMobileNetwork - A Platform for Providing Estimated Semantic Network Topology Data. *International Journal of Business Data Communications and Networking (IJBDCN)*, 9(4):46–64, October 2013.

[J2] M. von Hoffen and A. Uzun. Linked Open Data for Context-aware Services: Analysis, Classification and Context Data Discovery. *International Journal of Semantic Computing (IJSC)*, 8(4):389–413, December 2014.

Conference Proceedings

[C1] N. Bayer, D. Sivchenko, H.-J. Einsiedler, A. Roos, A. Uzun, S. Göndör, and A. Küpper. Energy Optimisation in Heterogeneous Multi-RAT Networks. In *Proceedings of the 15th International Conference on Intelligence in Next Generation Networks*, ICIN '11, pages 139–144, Berlin, Germany, October 2011. IEEE.

[C2] S. Dawoud, A. Uzun, S. Göndör, and A. Küpper. Optimizing the Power Consumption of Mobile Networks based on Traffic Prediction. In *Proceedings of the 38th Annual International Computers, Software & Applications Conference*, COMPSAC '14, pages 279–288, Los Alamitos, CA, USA, July 2014. IEEE Computer Society.

[C3] S. Göndör, A. Uzun, and A. Küpper. Towards a Dynamic Adaption of Capacity in Mobile Telephony Networks using Context Information. In *Proceedings of the 11th International Conference on ITS Telecommunications*, ITST '11, pages 606–612, St. Petersburg, Russia, August 2011. IEEE.

[C4] S. Göndör, A. Uzun, T. Rohrmann, J. Tan, and R. Henniges. Predicting User Mobility in Mobile Radio Networks to Proactively Anticipate Traffic Hotspots. In *Proceedings of the 6th International Conference on Mobile Wireless Middleware, Operating Systems, and Applications*, MOBILWARE '13, pages 29–38, Bologna, Italy, November 2013. IEEE.

[C5] E. Neidhardt, A. Uzun, U. Bareth, and A. Küpper. Estimating Locations and Coverage Areas of Mobile Network Cells based on Crowdsourced Data. In *Proceedings of the 6th Joint IFIP Wireless and Mobile Networking Conference*, WMNC '13, pages 1–8, Dubai, United Arab Emirates, April 2013. IEEE.

[C6] A. Uzun. Linked Crowdsourced Data - Enabling Location Analytics in the Linking Open Data Cloud. In *Proceedings of the IEEE 9th International Conference on Semantic Computing*, ICSC '15, pages 40–48, Los Alamitos, CA, USA, February 2015. IEEE Computer Society.

[C7] A. Uzun and A. Küpper. OpenMobileNetwork - Extending the Web of Data by a Dataset for Mobile Networks and Devices. In *Proceedings of the 8th International Conference on Semantic Systems*, I-SEMANTICS'12, pages 17–24, New York, NY, USA, September 2012. ACM.

[C8] A. Uzun, L. Lehmann, T. Geismar, and A. Küpper. Turning the OpenMobileNetwork into a Live Crowdsourcing Platform for Semantic Context-aware Services. In *Proceedings of the 9th International Conference on Semantic Systems*, I-SEMANTICS'13, pages 89–96, New York, NY, USA, September 2013. ACM.

[C9] A. Uzun, M. Salem, and A. Küpper. Semantic Positioning - An Innovative Approach for Providing Location-based Services based on the Web of Data. In *Proceedings of the IEEE 7th International Conference on Semantic Computing*, ICSC '13, pages 268–273, Los Alamitos, CA, USA, September 2013. IEEE Computer Society.

[C10] A. Uzun, M. Salem, and A. Küpper. Exploiting Location Semantics for Realizing Cross-referencing Proactive Location-based Services. In *Proceedings of the IEEE 8th International Conference on Semantic Computing*, ICSC '14, pages 76–83, Los Alamitos, CA, USA, June 2014. IEEE Computer Society.

[C11] A. Uzun, M. von Hoffen, and A. Küpper. Enabling Semantically Enriched Data Analytics by Leveraging Topology-based Mobile Network Context Ontologies. In *Proceedings of the 4th International Conference on Web Intelligence, Mining and Semantics*, WIMS '14, pages 35:1–35:6, New York, NY, USA, June 2014. ACM.

[C12] M. von Hoffen, A. Uzun, and A. Küpper. Analyzing the Applicability of the Linking Open Data Cloud for Context-aware Services. In *Proceedings of the IEEE 8th International Conference on Semantic Computing*, ICSC '14, pages 159–166, Los Alamitos, CA, USA, June 2014. IEEE Computer Society.

Contents

Part II Contribution

Acronyms

ABox	Assertional Box
AP	Access Point
API	Application Programming Interface
AuC	Authentication Center
B2B	Business-to-Business
B2C	Business-to-Consumer
BCCH	Broadcast Common Control Channel
BSS	Base Station Subsystem
BSSID	Basic Service Set Identification
BSC	Base Station Controller
BTS	Base Transceiver Station
CAS	Context-aware Service
CDMA	Code Division Multiple Access
CDR	Call Detail Records
CGF	Charging Gateway Function
CMO	Context Meta Ontology
CMOD	Context Meta Ontology Directory
CN	Core Network
CS	Circuit Switched
CSS	Cascading Style Sheets
CSV	Comma-separated Values
CDCApp	Context Data Cloud for Android App
DCS	Distributed Communication Sphere
EDGE	Enhanced Data Rates for GSM Evolution
EIR	Equipment Identity Register
eNB	eNode-B
EPC	Evolved Packet Core
FDMA	Frequency Division Multiple Access
FIPA	Foundation for Intelligent Physical Agents
FML	Framework Measurement Location

GeoRSS	Geographically Encoded Objects for RSS feeds
GERAN	GSM/EDGE Radio Access Network
GGSN	Gateway GPRS Support Node
GML	Geography Markup Language
GMSC	Gateway Mobile Switching Center
GPRS	General Packet Radio Service
GPS	Global Positioning System
GSM	Global System for Mobile Communications
HLR	Home Location Register
HSPA	High Speed Packet Access
HTML	Hypertext Markup Language
HTTP	Hypertext Transfer Protocol
IMEI	International Mobile Equipment Identity
IMSI	International Mobile Subscriber Identity
IoT	Internet of Things
IP	Internet Protocol
IRI	Internationalized Resource Identifier
ISDN	Integrated Services Digital Network
IT	Information Technology
ITU	International Telecommunication Union
JS	JavaScript
JSON	JavaScript Object Notation
LA	Location Area
LAC	Location Area Code
LBS	Location-based Service
LBG	Location-based Game
LCD	Linked Crowdsourced Data
LGD	LinkedGeoData
MDM	Measurement Data Manager
MCC	Mobile Country Code
MME	Mobility Management Entity
MNC	Mobile Network Code
MSC	Mobile Switching Center
MSIN	Mobile Subscriber Identification Number
NB	Node B
OGC	Open Geospatial Consortium
OMN	OpenMobileNetwork
OMNApp	OpenMobileNetwork for Android App
OMNG	OpenMobileNetwork Geocoder
OSM	OpenStreetMap
OTT	Over-the-top
OWL	Web Ontology Language
OS	Operating System
PDN-GW	Packet Data Network Gateway
PDP	Packet Data Protocol

PE	Positioning Enabler
PEM	Position Estimation Manager
PHP	Hypertext Preprocessor
POI	Point of Interest
PS	Packet Switched
QoC	Quality of Context
QoE	Quality of Experience
QoS	Quality of Service
RA	Routing Area
RAN	Radio Access Network
RDB	Relational Database
RDF	Resource Description Framework
RDFS	RDF Schema
RNC	Radio Network Controller
RSSI	Received Signal Strength Indicator
S-GW	Serving Gateway
SGSN	Serving GPRS Support Node
SMS	Short Message Service
SPARQL	SPARQL Protocol And RDF Query Language
SSID	Service Set Identifier
SWIG	Semantic Web Interest Group
TA	Tracking Area
TBox	Terminological Box
TDMA	Time Division Multiple Access
TSDO	Telecommunications Service Domain Ontology
Turtle	Terse RDF Triple Language
UML	Unified Modeling Language
URI	Uniform Resource Identifier
URL	Uniform Resource Locator
URA	UTRAN Registration Area
UTRAN	UMTS Terrestrial Radio Access Network
VLR	Visitor Location Register
VoID	Vocabulary of Interlinked Datasets
W3C	World Wide Web Consortium
WGS84	World Geodetic System 1984
WLAN	Wireless Local Area Network
WKT	Well-known Text
WWW	World Wide Web
XML	Extensible Markup Language

List of Figures

List of Tables

Zusammenfassung

Linked Data beschreibt Prinzipien zur Beschreibung, Veröffentlichung und Vernetzung strukturierter Daten im Web. Durch die Anwendung dieser Prinzipien entstand über die Zeit ein umfangreicher Graph von vernetzten Daten, welcher unter dem Namen *LOD Cloud* bekannt ist. Mobilfunkanbieter können von diesen Konzepten besonders profitieren, um eine größtmögliche Verwertung ihrer Mobilfunknetzdaten zu ermöglichen. Durch eine semantische Anreicherung ihrer Daten nach den *Linked Data* Prinzipien und der Verknüpfung dieser mit weiteren verfügbaren Informationsquellen in der *LOD Cloud*, können sie in die Lage versetzt werden, ihren Kunden innovative kontextbasierte Dienste anzubieten.

Für die Bereitstellung von semantisch angereicherten und kontextbasierten Diensten ist eine semantische und topologische Repräsentation von Mobilfunk- und WLAN-Netzen in Kombination mit einer Verknüpfung zu anderen Datenquellen unabdingbar. Diese Repräsentation muss die Standorte und Abdeckungsbereiche von Mobilfunkzellen und WLAN Access Points sowie deren Nachbarschaftsbeziehungen umfassen und diese auf geographische Bereiche unter Berücksichtigung ortsabhängiger Kontextinformationen abbilden. Zusätzlich sollte diese Beschreibung auch dynamische Netzinformationen, wie z.B. den Datenverkehr in einer Mobilfunkzelle, mit einbeziehen.

Zu diesem Zweck wird das *OpenMobileNetwork* als Kernbeitrag dieser Dissertation präsentiert, das eine Plattform zur Bereitstellung von approximierten und semantisch angereicherten Topologiedaten für Mobilfunknetze und WLAN Access Points in Form von *Linked Data* ist. Die Grundlage für das semantische Modell bildet die *OpenMobileNetwork Ontology*, die aus einer Menge von sowohl statischen als auch dynamischen *Network Context* Facetten besteht. Der Datenbestand ist zudem mit relevanten Datenquellen aus der *LOD Cloud* verlinkt. Einen weitereren Beitrag leistet die Arbeit durch die Bereitstellung von *Linked Crowdsourced Data* und der dazugehörigen *Context Data Cloud Ontology*. Dieser Datensatz reichert statische Ortsdaten mit dynamischen Kontextinformationen an und verknüpft sie zudem mit den Mobilfunknetzdaten im *OpenMobileNetwork*.

Verschiedene Applikationsszenarien und exemplarisch umgesetzte Dienste heben den Mehrwert dieser Arbeit hervor, der zudem anhand zweier separater Evaluationen untermauert wird. Da die Nutzbarkeit der angebotenen Dienste stark von der Qualität der approximierten Mobilfunknetztopologien im *OpenMobileNetwork* abhängt, werden die berechneten Mobilfunkzellen hinsichtlich ihrer Position im Vergleich zu den echten Standorten analysiert. Das Ergebnis zeigt eine hohe Qualität der Approximation auf. Bei den exemplarischen Diensten wird die Präzision des *Semantic Tracking* Dienstes sowie die Leistung des *Semantic Geocoding* Ansatzes evaluiert, die wiederum den Mehrwert semantisch angereicherter Mobilfunknetzdaten darlegen.

Abstract

Linked Data defines a concept for publishing data in a structured form with well-defined semantics and for relating this information to other datasets in the Web. Out of this concept, a huge graph of interlinked data has evolved over time, which is also known as the *LOD Cloud*. The telecommunications domain can highly benefit from the principles of *Linked Data* as a step forward for exploiting their valuable asset—the mobile network data. By semantically enriching mobile network data according to those principles and correlating this data with the extensive pool of context information within the *LOD Cloud*, network providers might become capable of providing innovative context-aware services to their customers.

Semantically enriched context-aware services in the telecommunications domain require a semantic as well as topological description of mobile and WiFi networks in combination with interlinks to diverse context sources. This description must incorporate the positions of mobile network cells and WiFi access points, their coverage areas, and neighbor relations, along with *dynamic network context* data (e.g., the generated traffic in a cell). In addition, *third-party context* sources need to be integrated providing location-dependent information such as popular points of interest visited during certain weather conditions.

The core contribution of this thesis is the *OpenMobileNetwork*, which is a platform for providing estimated and semantically enriched mobile and WiFi network topology data based on the principles of *Linked Data*. It is based on the *OpenMobileNetwork Ontology* consisting of a set of network context ontology facets that describe mobile network cells as well as WiFi access points from a topological perspective and geographically relate their coverage areas to other context sources. As another contribution, this thesis also presents *Linked Crowdsourced Data* and its corresponding *Context Data Cloud Ontology*, which is a crowdsourced dataset combining static location data with dynamic context information. This dataset is also interlinked with the *OpenMobileNetwork*.

Various application scenarios and proof of concept services are introduced in order to showcase the added value of this work. In addition, two separate evaluations are performed. Due to the fact that the usability of the provided services closely depends on the quality of the approximated network topologies, a distance comparison is performed between the estimated positions for mobile network cells within the *OpenMobileNetwork* and a small set of real cell positions. The results prove that context-aware services based on the *OpenMobileNetwork* rely on a solid and accurate network topology dataset. Concerning our proof of concept services, the positioning accuracy of the *Semantic Tracking* approach and the performance of our *Semantic Geocoding* are evaluated verifying the applicability and added value of semantically enriched mobile and WiFi network data.

Part I
Basics

Chapter 1
Introduction

In the last decade, the *World Wide Web (WWW)* has more and more moved from a *Web of Documents* towards a *Web of Data*.[1] The concept of *Linked Data* [62] took the driving force position in pushing this data-oriented change within the Web. According to this concept and its four basic rules defined by Berners–Lee [20], information is published in a structured form with well-defined meaning and related to other datasets in the Web building a huge graph of interlinked data, which is also known as the *Linking Open Data (LOD) Cloud* [2]. As this global Web of semantically enriched and machine-readable data met Berners–Lee's initial idea when he proclaimed his vision of the *Semantic Web* [18, 21], he referred to this approach as the "Semantic Web done right" [19].

A major advantage of *Linked Data* is that it enables the retrieval of data within the *LOD Cloud* in a uniform manner. In contrast to other external data sources (e.g., Web Application Programming Interfaces (API)) where developers are forced to implement against proprietary APIs with different data formats and parse the returned results to the desired data model, the *LOD Cloud* provides data in a uniform format (namely the *Resource Description Framework (RDF)* [85, 121]) that can be queried via the standardized SPARQL interface [139]. This process reduces the manual implementation effort for extracting information to a great extent.

The huge graph of semantically enriched and (more or less) publicly available data as well as the uniform data model makes this concept very interesting and relevant for various application areas and companies of different domains. Prominent examples are *BBC* [122] reimplementing their sports and online presence as well as *Financial Times* [123] enhancing their online and print product portfolio with semantic technologies.

One specific candidate, which can highly benefit from the principles of *Linked Data* in general and specifically from the data within the *LOD Cloud*, is the class of *Context-aware Services (CASs)*. These services adapt their behavior or the provided content [29] based on the (current) contextual situation of an entity that could be a

[1]https://www.w3.org/standards/semanticweb/.

A. Uzun, *Semantic Modeling and Enrichment of Mobile and WiFi Network Data*, T-Labs Series in Telecommunication Services,
https://doi.org/10.1007/978-3-319-90769-7_1

person, a place, or an object [46]. In order to derive the contextual situation of an entity, heterogeneous sources and various forms of context information are utilized and correlated for which the extensive pool of available context data within the *LOD Cloud* is predestined for.

In parallel to the developments in the Web, the telecommunications domain has also witnessed a shift over the last decade. *Over-the-top (OTT)* service providers, such as *Google*[2] or *Facebook*,[3] have more and more taken a front role in the Information Technology (IT) market achieving huge revenues with their services, whereas telecommunication providers still struggle with their position as a bit pipe being the provider and maintainer of the fixed line and mobile networks on top of which those services operate [12]. Telecommunication providers mainly put their focus on network provisioning, planning, and maintenance issues. However, especially in the era of ubiquitous mobile devices and increasing mobile Internet usage, mobile network data (e.g., user movements, number of users and traffic produced in mobile network cells, service usage information, or smartphone capabilities) turns into a valuable asset that can be utilized for establishing mobile network operators as service enablers who become capable of providing diverse context-aware services to their customers in the *In-house*, *Business-to-Customer (B2C)* as well as *Business-to-Business (B2B)* application domains.

In-house services can be utilized by mobile network operators for optimizing their internal processes and infrastructure based on certain circumstances. By doing so, they can reduce the overall costs or improve the *Quality of Service (QoS)* and the user-perceived *Quality of Experience (QoE)* of their networks. One example is the development of a power management in mobile networks, in which the current state of the network is analyzed in order to de- and reactivate mobile network cells based on network usage profiles. This leads to energy as well as cost savings.

B2C services, on the other hand, comprise applications based on mobile network data that telecommunication providers are potentially able to provide to their end customers. All kinds of *Location-based Services (LBSs)* as well as positioning methods, for instance, can be part of this group. Furthermore, through interfaces and dashboards, mobile network operators can provide analytics information to third-parties in the form of B2B services. Stores interested in the age groups of people passing by, for example, can get this information in order to adjust their product portfolio and marketing campaigns accordingly.

Meanwhile, some network providers have realized the relevance of this opportunity and started to launch products or startups that apply (outdoor and indoor) analytics on their own network data in order to provide services focusing on the B2B market. *Telefónica Smart Steps* [15], for example, is such a product that analyzes crowd movements out of mobile network data providing footfall information to businesses for a specific location. *MotionLogic*[4] is a venture of *Deutsche Telekom*

[2]http://www.google.com/.

[3]http://www.facebook.com/.

[4]http://www.motionlogic.de/.

and goes into a similar direction by selling "geomarketing insights" based on "anonymous signalling data".

Nevertheless, the full potential of network data exploitation is not achieved yet. None of these approaches consider the integration of semantic technologies (as described above) into their services. Enriching mobile network data with well-defined and highly expressive semantics and correlating this data with heterogeneous context data sources (available within the *LOD Cloud*, for example) will enable new application areas for mobile network operators and foster innovative context-aware services. This fusion will facilitate the usage of sophisticated semantic queries that do not (only) rely on network parameters, geofences, or geo coordinates, but rather on semantic relations between different network components and external context data sources. In combination with powerful reasoners[5] that generate implicit knowledge out of the semantically enriched data, network providers might be given a head start compared to other service providers, which do not possess these assets.

Another drawback of the provided solutions stems from the isolated view on the data. Telecommunication providers are only able to provide services based on their own data or - in some cases - even only to their own customers. A global view on the provided analytics information comprising not only data of a specific operator's own network, but rather of all networks, is not possible. In addition, a potential LBS or a positioning solution, for example, can only be used by the customers of this operator. Integrating data of other networks including also (private) WiFi access points (AP), however, might enhance the quality of the services or extend the potential user base.

Moreover, having a look from the perspective of the *Web of Data*, mobile network data that is theoretically made available for the public in *Linked Data* format will add a new dimension to the *LOD Cloud* allowing users worldwide to interlink their own data to mobile networks and come up with new services.

1.1 Problem Statement and Research Questions

The provision of semantically enriched context-aware services in the In-house, B2C as well as B2B application areas requires a semantic description of mobile and WiFi network data as well as a geographic and topological view on the respective network components in combination with interlinks to diverse context sources.

Taking the In-house power management scenario as an example, in order to de- or reactivate a mobile network cell for energy saving purposes, *static network context* information is needed including the technical capabilities of the base station (e.g., its mobile network generation), but also the geographic position and coverage area of the cell as well as neighboring cells surrounding it. Furthermore, *dynamic network context* data that describes constantly changing information within the network components (e.g., the generated traffic or the number of users in a cell), has to be known. Finally, external *third-party context* sources need to be integrated that pro-

[5]https://en.wikipedia.org/wiki/Semantic_reasoner#List_of_semantic_reasoners.

vide location-dependent context information in the coverage area of the cell such as popular points of interest (POI) or events visited during certain weather conditions or on holidays, for example.

Existing ontologies and models typically concentrate on static network context in terms of technical capabilities describing aspects like network connectivity, mobile devices, or a combination of both [8, 53, 150]. A variety of them focus on different capabilities of mobile devices and comprise information about the supported communication standards or device-specific characteristics such as the resolution of the display, the operating system, the battery capacity, or processing power. Qiao et al. [120, 156], on the other hand, introduces concepts and features of mobile as well as fixed networks and models the relationship between different network access technology types.

However, these vocabularies do not incorporate a geographic and topological view on mobile and WiFi networks. They do not model concepts for describing the position of a mobile network cell or a WiFi access point, their coverage areas, or neighbor relations, nor do they take dynamic network context information, such as the services used within a cell, into consideration.

Besides a model on the concept-level, another problem exists in terms of the availability of worldwide mobile network topology data as telecommunication providers keep their asset very secret. Over time, commercial as well as open data project providers came up collecting cell and WiFi AP information via crowdsourcing and estimating their positions. Two famous open data projects are *OpenCellID*[6] and *OpenBMap*[7] that allow access to their entire dataset.

By the time of our studies, both projects lacked relevant network information necessary for the realization of the above mentioned application scenario. In addition, dynamic network context was not considered at all and no quality evaluation of the data was available. This, however, is of utter importance as the QoE of the context-aware services is strongly related to the usage of accurately mapped mobile network topologies.

The complexity and power of semantically enriched services further depends on the integrated third-party context. A correlation of network topology data with available context sources in the *LOD Cloud*, such as *LinkedGeoData (LGD)*[8] [144], will already enable the development of basic context-aware services. Such a service within the power management scenario will allow semantic queries to retrieve all mobile network cells with a certain number of users covering a specific POI, for instance.

Nevertheless, the variety of realizable scenarios is still limited since geo-related datasets in the *LOD Cloud* are rather of static nature and mainly consist of information, such as a name, geo coordinates, an address, or opening hours, for a place. They do not provide dynamic characteristics about certain places such as its popularity, the "visiting frequency" (determined by the number of check-ins, for example) or dwell

---

[6]http://www.opencellid.org/.

[7]http://www.openbmap.org/.

[8]http://www.linkedgeodata.org/.

time of users in specific contextual situations (e.g., for certain weather conditions or on holidays).

Commercial providers, such as *Foursquare*[9], *Google*, or *Facebook*, possess high quality location (and check-in) data fulfilling some of the mentioned aspects, which can also be requested through their APIs. However, a deeper look at their terms of use[10,11] reveals that storing this data in another database, combining or modifying it, is not allowed. In addition, these datasets are neither available in RDF format nor published as *Linked Data* restricting their use within the *LOD Cloud*.

For enabling powerful and semantically enriched context-aware services in the In-house, B2C as well as B2B application domains, linked datasets and corresponding ontologies are required that provide semantically enriched network topology data and extend static location data with dynamic context information. This thesis tackles these challenges and gives answers to the following research questions derived from the problem statement:

1. How to model an ontology that incorporates a geographic and topological view on mobile as well as WiFi networks and further takes dynamic network context information into consideration?
2. How to collect and accurately estimate worldwide mobile and WiFi network data that suffices the requirements for context-aware services in the In-house, B2C as well as B2B application areas?
3. How to interlink the semantically enriched network topology data to diverse context sources?
4. How to create a dataset and its corresponding vocabulary that extends static location data with dynamic context information?
5. How to enable network operators to leverage semantically enriched mobile and WiFi network data for providing innovative context-aware services?

1.2 Contribution

The core contribution of this work is the *OpenMobileNetwork (OMN)*[12] [C7, J1], which is a platform for providing approximated and semantically enriched mobile network and WiFi access point topology data based on the principles of *Linked Data* [6]. Since mobile network operators keep their asset (being the network topology) very secret, network measurements are constantly collected via a crowdsourcing approach [C8] in order to infer the topology of mobile networks and WiFi access points worldwide [C5, J1].

---

[9]http://www.foursquare.com/.

[10]https://developer.foursquare.com/overview/venues.

[11]https://developers.google.com/maps/terms.

[12]http://www.openmobilenetwork.org/.

The foundation of the provided dataset within this platform is the *OpenMobileNetwork Ontology*[13] [C11] consisting of a set of static and dynamic network context ontology facets that describe mobile radio access networks (RAN) and WiFi access points from a topological perspective and geographically relate the coverage areas of these network components to each other.

Additional context information about the locations covered by those reception areas is acquired through interlinks to several datasets. In addition to well-known geospatial datasets, this thesis presents *Linked Crowdsourced Data (LCD)* [C6] and its corresponding *Context Data Cloud Ontology*[14] as another contribution, which is a crowdsourced dataset linking dynamic parameters (e.g., check-ins, ratings, or comments), specific context situations (e.g., weather conditions, holiday information, or measured networks) as well as additional domain-specific information (e.g., dishes of a restaurant) to static location data enabling the development of sophisticated context-aware services.

An applicability study is done by demonstrating various semantic In-house, B2B as well as B2C application scenarios and highlighting their added value within this work. The In-house service example incorporates a power management in mobile networks[15] [C11] for energy saving and overall cost reduction, in which the current state of the network is analyzed in order to automatically de- and reactivate mobile network cells based on network usage profiles. The exemplary B2C services, on the other hand, comprise several *Semantic Positioning* solutions, such as a *Semantic Tracking* [C9, C10] and a *Semantic Geocoding*, that overcome the limitations of classic geocoding as well as geofencing methods and add semantic features to location-based services, while the usage of semantically enriched location analytics is presented as an example for a B2B service [C6].

In order to ultimately determine whether semantically enriched mobile network data can be utilized by network operators for providing innovative context-aware services, two separate evaluations are performed that focus on the added value gained through the services. Due to the fact that the usability of the provided services closely depends on the quality of the estimated network topologies and the accurate geographic mapping of the networks, a distance comparison between the estimated positions for mobile network cells within the *OpenMobileNetwork* and a small set of real cell positions (provided by an operator) is performed. In addition, the positioning accuracy of the *Semantic Tracking* approach as well as the performance of our *Semantic Geocoding* are evaluated highlighting the applicability and added value of semantically enriched mobile and WiFi network data.

---

[13] *omn-owl*, http://www.openmobilenetwork.org/ontology/.

[14] *cdc-owl*, http://www.contextdatacloud.org/ontology/.

[15] http://www.openmobilenetwork.org/comgreen/.

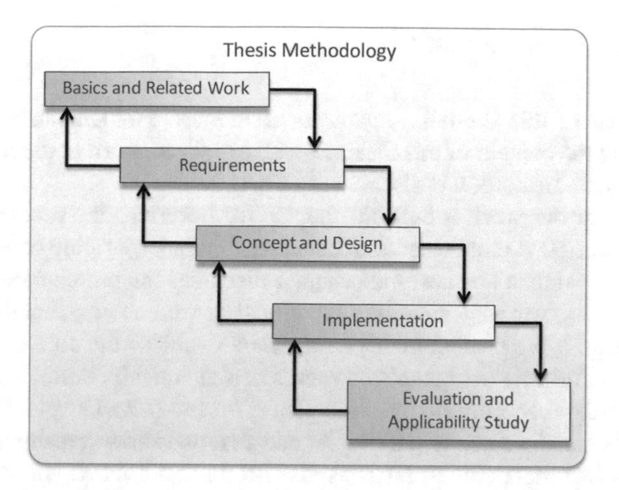

Fig. 1.1 Thesis Methodology

1.3 Methodology

The work within this thesis was done following a classic waterfall model [127] used in software development projects, which is depicted in Fig. 1.1. Each phase of the entire research and development process was separated into work packages, where the preceding work package was the input for the next phase.

The first work package comprised the acquirement of fundamental technologies and methods as well as a detailed analysis of related work in the relevant research fields. Having an understanding of the state of the art, the requirements for our contribution were specified in the second work package. This specification included the application areas in question, the context data requirements as well as the functional and non-functional requirements of the prospective platform. Based on these requirements, the concept and design process of the *OpenMobileNetwork* as well as *Linked Crowdsourced Data* was executed. This design led to the implementation, which is part of the fourth work package. The developed proof of concept of the platform as well as the service demonstrators enabled us to perform the evaluation and applicability study.

Various improvements were made to the *OpenMobileNetwork* throughout the time due to new application scenarios or requirements defined by new research projects, which is why we had several cycles in the waterfall model between the related work and evaluation work packages. Furthermore, the development of the *OpenMobileNetwork* as well as *Linked Crowdsourced Data* happened in time-delayed, but parallel work streams.

1.4 Thesis Outline and Structure

The remainder of this thesis is organized as follows: Fundamentals required for understanding the content of this thesis as well as related work in the research areas affected by our contribution is discussed in Chap. 2.

The basis for our work is built in Chap. 3 that describes the requirements for a semantically enriched mobile network data platform incorporating context information from external data sources. This chapter discusses the prerequisites on context data and gives an overview about the functional as well as non-functional requirements. In Chap. 4, we present the *OpenMobileNetwork* as the core contribution of the thesis and illustrate the complete process of semantically enriching mobile and WiFi network data according to the principles of *Linked Data*. Chapter 5, on the other hand, extends the aforementioned chapter and describes how semantically enriched network topology data can be interlinked with diverse context sources. Here, we introduce *Linked Crowdsourced Data* as another contribution of the work and highlight its added value in comparison to existing geo-related datasets. Insight into the implementation of the *OpenMobileNetwork* is given in Chap. 6 that provides an end-to-end view of all platform components incorporating the system architecture, the smartphone clients as well as the *OpenMobileNetwork* backend. The contribution is finished with Chaps. 7 and 8. Chapter 7 showcases the added value of the *OpenMobileNetwork* by introducing several In-house, B2C, and B2B services, whereas the corresponding proof of concept demonstrators are presented in Chap. 8.

Chapters 9 and 10 comprise our evaluation. The main goal of the evaluation is to highlight the added value of semantically enriched mobile network data that is interlinked to other datasets in the *LOD Cloud* for the development of context-aware services. Due to the fact that the performance of these services (in terms of accuracy and user experience) strongly depends on the preciseness of the approximated network topology, a quality analysis of the *OpenMobileNetwork* data is conducted in Chap. 9, which is based on a distance comparison of the estimated mobile network cells within the *OpenMobileNetwork* to the real positions of these cells. In Chap. 10, on the other hand, we focus on the applicability of the services that have been introduced within this doctoral thesis and illustrate how semantically enriched mobile and WiFi network data improves classic geofencing and geocoding solutions.

The doctoral thesis is concluded with Chap. 11 summarizing the contribution and discussing the research results. Chapter 12 provides an outlook on possible future research.

Chapter 2
Basics and Related Work

Chapter 2 discusses the fundamentals required for understanding the content of this thesis and further highlights related work in the research areas affected by our contribution. As our work is built on mobile network data that is semantically modeled from a topological perspective, we briefly introduce mobile networks in their different generations in Sect. 2.1. Section 2.2, on the other hand, makes a deep dive into context-awareness by defining the notion of context and explaining the complete context management process. In Sect. 2.3, we describe key technologies and approaches of the *Semantic Web* that are mainly applied in our work. Related platforms and datasets are discussed in Sect. 2.4, whereas Sect. 2.5 provides a look into related context ontologies. Additional related work, which is analyzed for specific parts of our concept, is directly illustrated within the concept chapters of our contribution.

2.1 Mobile Networks

The dataset of the *OpenMobileNetwork* comprises worldwide network topology data of different mobile network generations. With the help of the corresponding *OpenMobileNetwork Ontology*, mobile radio access networks are described from a topological perspective consisting of radio cells, their positions and coverage areas, and information about neighboring cells. In order to create an understanding for the mobile network topology, we briefly introduce mobile networks based on the book by Sauter [132] and primarily focus on access network components. Due to the fact that our contribution specifically targets a semantic representation of the topology, we do not discuss the concepts of mobility management, such as handover or location management, in this section. Please refer to [132] for a detailed understanding of these aspects.

A. Uzun, *Semantic Modeling and Enrichment of Mobile and WiFi Network Data*, T-Labs Series in Telecommunication Services,
https://doi.org/10.1007/978-3-319-90769-7_2

In general, mobile networks - irrespective of their generation - are based on an infrastructure that consists of a *radio access* and a *core network (CN)*. The radio access network comprises a number of base stations and other supporting nodes that enable a wireless connection of a mobile device to the network. A base station is a transceiver component that covers a limited geographic area, which is called a radio cell. Depending on the mobile network, the regional deployment, and the required capacity, the coverage area of a cell differs in its radius.

The core network, on the other hand, represents the backbone of the mobile network. It interconnects several access networks and performs key network functions such as switching calls, mobility management, or subscriber management. Depending on the mobile network technology, the core network is either based on a *circuit-switched (CS)* or *packet-switched (PS)* system. A circuit-switched network mainly serves for voice communication and establishes a direct connection between two calling parties, whereas a packet-switched network is designed for *Internet Protocol (IP)* data packet transmission with no logical end-to-end connection.

Figure 2.1 provides an overview of the available mobile networks including their components and illustrates the interaction of them. The lines in between the components represent on which plane the communication takes place. The *user plane* (highlighted as a straight line) incorporates all channels, protocols as well as methods for transmitting user data (e.g., voice or Internet). Signaling data, such as call/data session setups, handovers, location updates, or paging, is processed in the *control plane* (marked as a dotted line).

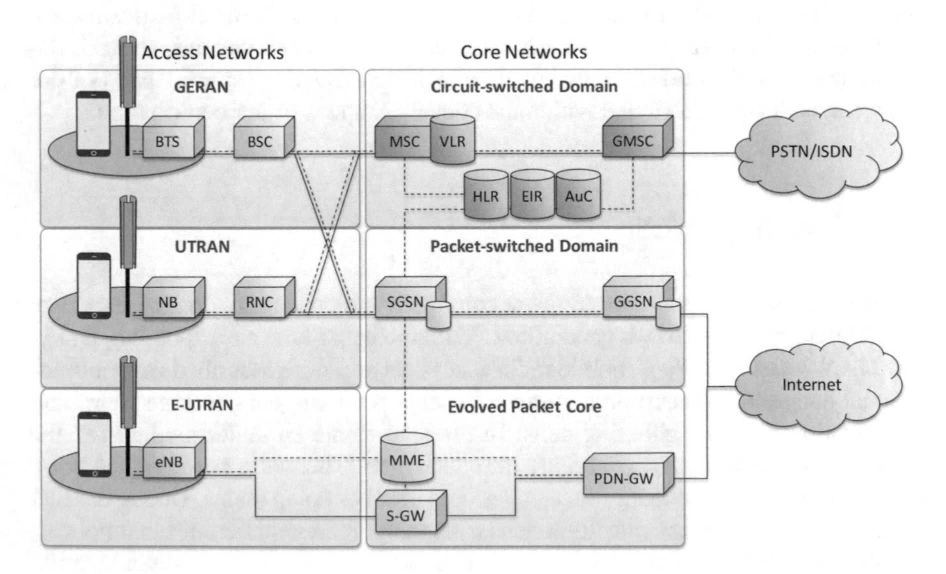

Fig. 2.1 Mobile networks

2.1.1 Global System for Mobile Communications (GSM)

After the first generation of analog systems, the *Global System for Mobile Communications (GSM)* was the first standard of the second generation of mobile networks (i.e., 2G) that was specified by the *European Telecommunications Standards Institute (ETSI)*.[1] Originally, it was designed as a digital circuit-switched network to enable voice communication. Later, the standard was enhanced by the *General Packet Radio Service (GPRS)* and the *Enhanced Data Rates for GSM Evolution (EDGE)* (known as 2.5G) accompanied by an additional core network to also allow packet-switched data transmission.

A number of *Base Station Subsystems (BSS)* form the *GSM/EDGE Radio Access Network (GERAN)*, where each BSS consists of multiple *Base Transceiver Stations (BTS)* and a *Base Station Controller (BSC)* that controls and coordinates the BTS by reserving radio frequencies or managing handovers in the same BSS, for example.

The BTS establishes the connection between the mobile device and the network. It is theoretically capable of covering a geographic area of up to 35 km. However, in residential and business areas, the radius of a cell coverage area is adjusted to 3 or 4 km due to the limited number of subscribers it can serve at the same time, whereas highly frequented areas, such as shopping malls, are covered by cells with a radius of several 100 m. Rural areas usually consist of base stations with a coverage area of up to 15 km.

Usually, a base station is surrounded by a number of neighboring sites that have to communicate on different frequencies. This limits the number of frequencies available per base station and thus the provided capacity. A solution for reusing frequencies is achieved by splitting the coverage area of a base station to several sectors sending on different frequencies with a dedicated transmitter. By doing so, each sector defines its own cell with a unique *Cell-ID*. Several cells are further clustered into a *Location Area (LA)* for reducing the resource consumption when paging the mobile devices. Every LA is uniquely identified by a *Location Area Code (LAC)*.

Figure 2.2a illustrates base stations with neighboring sites forming radio cells that are clustered into *LAs*. In Fig. 2.2b, a possible sectoring of a cell is demonstrated. With slight differences in the design (e.g., *Routing Areas (RA)* for GPRS, *UTRAN Registration Areas (URA)* for UMTS, and *Tracking Areas (TA)* in LTE), the cell topology structure is the same in all mobile networks.

Communication between a mobile device and a BTS is based on a combination of the *Frequency Division Multiple Access (FDMA)* and the *Time Division Multiple Access (TDMA)* methods. The first channel access method separates the communicating users to different frequencies. In TDMA, on the other hand, each calling party is given a communication time slot of 577 μs (i.e., a burst) out of 8 within a time frame, which is repeating itself with every frame. Besides dedicated channels that handle the communication for active users, the *Broadcast Common Control Channel (BCCH)* broadcasts relevant cell information to all mobile devices within the cell.

[1] http://www.etsi.org/.

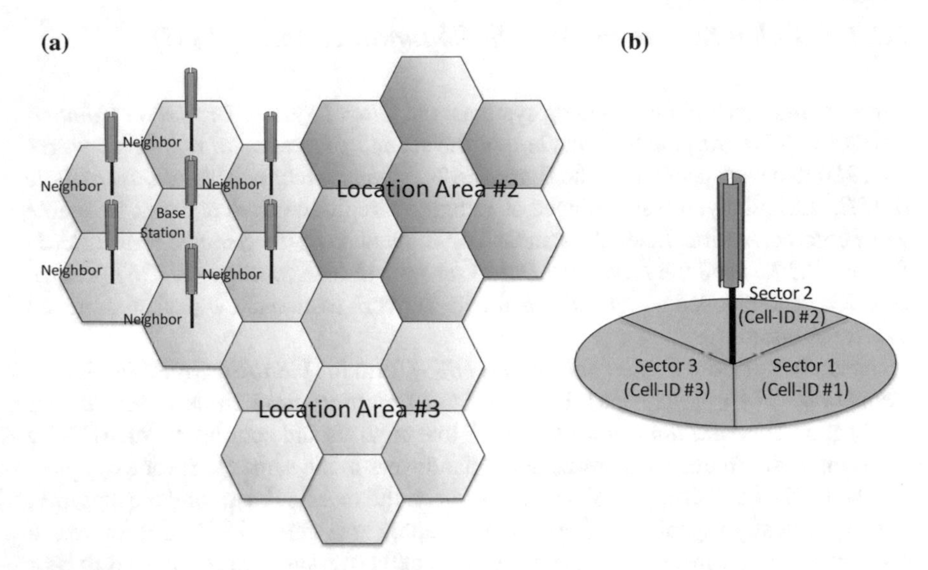

Fig. 2.2 Mobile network cell structure **a** Location areas incl. base stations with neighbors **b** Base station cell sectors

This information includes a list of neighboring cells, so that the mobile devices do not have to measure the whole frequency band for them. Furthermore, the *Mobile Country Code (MCC)* as well as the *Mobile Network Code (MNC)* of the cell is broadcasted along with the Cell-ID and the LAC.

The circuit-switched core domain of GSM comprises the *Mobile Switching Center (MSC)* as its central component that manages all connections between the subscribers, performs call routings, and is responsible for the mobility management of the users. Each MSC coordinates a number of BSSs and is accompanied by a *Visitor Location Register (VLR)*, which is a database for temporarily storing information about the subscribers currently being active in the coverage area of the MSC. This information is copied from the *Home Location Register (HLR)* that permanently saves all subscriber data.

A very important record within the HLR is the *International Mobile Subscriber Identity (IMSI)* that consists of three numerical elements for uniquely identifying a user. The 3-digit MCC stands for the home country of the user (and implicitly of the mobile network operator), whereas the MNC is a 2–3 digits number representing nation-wide telecommunication providers. Both codes are used in combination in order to uniquely determine a mobile network operator worldwide. The *Mobile Subscriber Identification Number (MSIN)*, on the other hand, consists of 10 digits and identifies the subscriber.

Two other components of the core domain are the *Equipment Identity Register (EIR)* and the *Authentication Center (AuC)*. While the former is a database for user device data, the latter is utilized for protecting user identity and data transmission. The transfer of voice data from the mobile network to the *Public Switched Telephone*

Network (PSTN) or to the *Integrated Services Digital Network (ISDN)* is enabled by the *Gateway Mobile Switching Center (GMSC)*. At the same time, this gateway also serves as the entry point to the mobile network from fixed line networks.

With the introduction of GPRS and later EDGE, an additional packet-switched core network was implemented for enabling packet-switched data transmission with data rates up to 220kbps. This core network consists of two main elements: The *Serving GPRS Support Node (SGSN)* is similar to the MSC. It establishes a connection between the radio access network and the packet-switched core domain and is responsible for tunneling user sessions to the *Gateway GPRS Support Node (GGSN)*, which enables mobile subscribers to get access to the Internet.

2.1.2 Universal Mobile Telecommunications System (UMTS)

The third mobile network generation (i.e., 3G) was introduced with the standardization of the *Universal Mobile Telecommunications System (UMTS)* by the 3rd *Generation Partnership Project (3GPP)*.[2] It supported data rates up to 384kbps in its early releases, while the deployment of the *High-Speed Packet Access (HSPA)* made data rates up to 7.2Mbps possible.

UMTS reuses the circuit-switched and packet-switched core network of GSM, but is based on an entirely new designed radio access network. Similar to GERAN, the *UMTS Terrestrial Radio Access Network (UTRAN)* consists of a set of *Radio Network Controllers (RNC)* - each being responsible for multiple *Node-Bs (NB)*. The NB is the counterpart to the BTS with the main difference in the applied channel access method. In contrast to FDMA and TDMA, a NB works with (a variation of the) *Code Division Multiple Access (CDMA)* method. This method makes it possible for a base station to communicate with many mobile devices at the same time on the same frequency. Here, each mobile device encodes the data to be sent with a special code pattern before transmission, which is then decoded by the base station since the codes of each user are known to it.

2.1.3 Long Term Evolution (LTE)

A completely new infrastructure for the radio access as well as core network was designed with the standardization of *Long Term Evolution (LTE)* by 3GPP, which forms the fourth generation of mobile networks (i.e., 4G). The major difference and advancement to the former generations is that it completely relies on IP-based protocols on all interfaces. In addition, it only consists of three components for the user plane reducing the complexity of the architecture.

---

[2]http://www.3gpp.org/.

The radio access network comprises only a single element, which is also the most complex component in the whole infrastructure. In contrast to GSM and UMTS, the *eNode-B (eNB)* is an autonomous unit taking over not only tasks of a "traditional" base station, but also functions of a BSC or RNC. By doing so, it is responsible for user management, for reserving air interface resources as well as for mobility management, among other things. It is further capable of performing handovers between eNBs.

Three components are part of the core network, which is also called the *Evolved Packet Core (EPC)*. The *Mobility Management Entity (MME)* is the managing unit of the network and is similar to the SGSN in its functions except for the fact that it does not handle user data, but only signaling data. This component tracks idle mobile devices in the TAs, supports the handover procedures of the eNBs, performs handovers to the GSM or UMTS network, coordinates the establishment of an IP tunnel between an eNB and a gateway to the Internet, and exchanges authentication information with a mobile device when it is attached to the network.

The *Serving Gateway (S-GW)*, on the other hand, tunnels user data packets between the eNBs and the *Packet Data Network Gateway (PDN-GW)*, which is the gateway to the Internet. It is expected that this gateway will replace the GGSN in future.

Please note that the dataset of the *OpenMobileNetwork* mainly consists of cell data for GSM and UMTS rather than LTE. This is due to the fact that by the time of systematically collecting data, smartphones supporting LTE were rather in the minority.

2.2 Context-awareness

In order to achieve a common understanding for *Context-awareness*, it is required to have a sound definition of the notion of *context* that fits to the scope of this thesis.

2.2.1 Definition of Context

Formulating a precise definition of context is somewhat challenging since several definitions and classifications have been proposed over the years that differ (slightly) based on the application scenarios they are used in. In a very early work, Schilit et al. [134], for instance, claim that context comprises the three relevant aspects "where you are, who you are with, and what resources are nearby" and list "lighting, noise level, network connectivity, communication costs, communication bandwidth, and even the social situation" as possible examples for context parameters in addition to the location of a user. Lenat [90], on the other hand, identifies twelve very fine-grained dimensions of context such as *time*, *type of time*, *geo location*, *culture*, or *granularity*, whereas a hierarchical classification of context is provided by Schmidt et. al [135] with *human factors* and the *physical environment* being at the top level.

A first survey about context in mobile computing is presented by Chen and Kotz [34] who classify context into four dimensions based on the work in [134], namely *computing context*, *user context*, *physical context*, and *time context*. For the mobile computing domain, they separate context into *active* and *passive context*. Active context has a direct impact on the behavior of an application, while passive context is classified as rather relevant, but not critical.

The most well-known definition of context is given by Dey [46] in his key article who describes context as "any information that can be used to characterize the situation of an entity. An entity is a person, place, or object that is considered relevant to the interaction between a user and an application, including the user and applications themselves."

Zimmermann et al. [164] analyze several context definitions and refine the one of Dey by a formal as well as operational extension. By doing so, the authors present the *location*, *time*, *activity*, *relations*, and *individuality* as five categories that describe the context of an arbitrary entity. They argue that "the activity predominantly determines the relevancy of context elements in specific situations, and the location and time primarily drive the creation of relations between entities and enable the exchange of context information among entities."

As we can see from existing literature, there are multiple definitions and classifications of the notion of context, which sometimes use different names for the same type of context or differ in their granularity or the applied application scenarios.

In this thesis, we do not make one more attempt for providing a general context definition (since the available definitions and classfiications fully fit to our case), but rather focus the term to our needs and examine specific related work. Due to the fact that our work is placed within the telecommunications domain, we concentrate on all kinds of context data that is related to mobile networks and that is utilizable within the potential application areas defined in Sect. 3.1. We define *network context* as the top-level class of context including static as well as dynamic network context and accompany this type of information with third-party context. The context of a user is either embedded into the network context in the form of the user location (determined by network components), profiles (as maintained by network providers), or service usage information, or it is part of the third-party context source in the form of location preferences, for instance. Please see Sect. 3.2 for a detailed classification of context data as used within this thesis. Therefore, referring to Dey's definition [46], we define context as follows:

> Context is any (internal and external) information that can be used to characterize the current state of a mobile network including the interaction between different network components.

Services that utilize (heterogeneous sources of) context information for adapting their behavior or the provided content [29] are known as *context-aware services*.

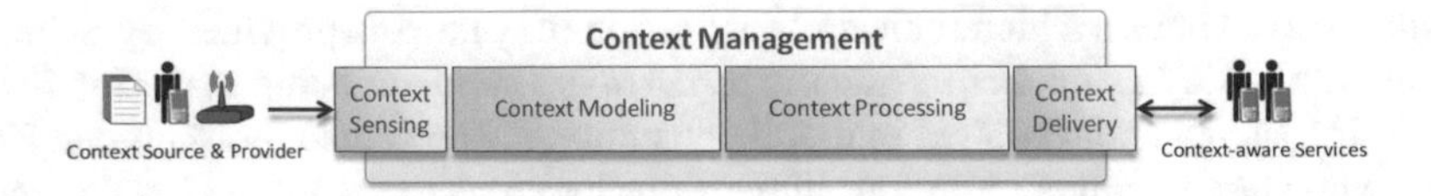

Fig. 2.3 Context management workflow

2.2.2 *Context Management*

After defining the notion of *context* and giving examples for certain types and classes of context information from existing literature, we briefly discuss the complete processing workflow of context data in this section, which is also known as *context provisioning* [67] or *context management*. This workflow comprises the steps of sensing (i.e., collecting), modeling, processing, delivering (i.e., disseminating), and consuming context information as exemplary illustrated in Fig. 2.3.

Our thesis contribution follows this process from a conceptual perspective, meaning that each part of our selected approach can be mapped to a certain step of the context management workflow. These aspects are specifically highlighted throughout the next subsections.

2.2.2.1 Context Sensing and Crowdsourcing

Context Sensing stands for the process of extracting and collecting relevant context information from diverse context sources. As the name implies, a *context source* is an entity, where context data originates from. Typical context sources are smartphones (see Sect. 4.3), network elements in the mobile network (see Sect. 4.1.1), sensors, Web APIs, or databases, for example. The role of the *context provider* is sometimes synonymously used to context source or is defined as the operator of it (e.g., a network operator) [67] offering interfaces in order to make the context data available for further processing [52, 98]. In Sect. 3.2.1, we analyze the potential of the *LOD Cloud* serving as a context provider with a huge set of heterogeneous context sources.

Extracting data from typical context sources, such as smartphones or Web APIs, is more or less trivial from an implementation point of view. Taking mobile devices as an example, developers need to implement a service against the sensor interfaces (e.g., the *Global Positioning System (GPS)* module) of a smartphone operating system (OS) in order to fetch the current status of the sensors integrated in the mobile device [C1]. However, for achieving realistic interpretations of macroscopic context situations, such as the current mobile network demand of a certain region or the average number of users visiting a certain POI at night times, context data needs to be collected on a large scale from a representative number of context sources, which turns out to be a very challenging task. One solution for this problem is the concept of *crowdsourcing* as introduced by Howe [69]. It describes an approach for sourcing a specific task to a motivated "crowd" in combination with an incentive, so that this crowd finds a solution to this problem. In contrast to "outsourcing", the task to be solved is not given to a specific group or a company, but rather to an undefined public [5].

According to Li et al. [91], crowdsourcing has the advantage that tasks can be completed at lower cost compared to when working on them in-house. Furthermore, they can be completed much faster (with the support of a "global workforce") since tasks can be separated into sub-tasks allowing people to work on them in parallel. However, the major challenge in applying a crowdsourcing approach is to find strategies on how to motivate the crowd. In his study, Hossain [68] summarizes motivation as either being intrinsic or extrinsic. Intrinsic motivation can be pushed by making the task itself attractive and enjoyable to work on, whereas extrinsic motivation can be increased by using *financial motivators* (e.g., financial benefits), *social motivators* (e.g., peer recognition), or *organizational motivators* (e.g., career development).

One of the famous examples for exploiting large-scale context information from smartphones collected via crowdsourcing is the car traffic information highlighted in the *Google Maps*[3] app [145]. In return for using the app with no cost, each *Android*[4] phone owner delivers valuable location information to *Google* out of which car traffic predictions are calculated. Similar to our work, Faggiani et al. [50] present another smartphone-based crowdsourcing approach and collect measurements from mobile devices for network monitoring purposes. By doing so, various application scenarios, such as signal coverage maps, can be built in order to enable network operators to optimize their networks in areas with poor or fluctuating signal quality.

We apply the concept of crowdsourcing for acquiring static as well as dynamic network context data from smartphones in order to approximate mobile network topologies worldwide. In addition, we collect third-party context in the form of context-specific user location preferences for correlating this data with the estimated network topologies. By doing so, we provide intrinsic incentives (to research communities, for example) as our work comprises open datasets as part of the *LOD Cloud* that can be utilized by anybody. Furthermore, we make use of gamification and community aspects in order to motivate users extrinsically. Please see Sects. 4.3 and 5.2.1 for further details.

2.2.2.2 Context Modeling and Processing

Data originating from a context source is often initially available in its raw form (referred to as *low-level context information* [67]) and requires additional processing steps in order create *higher-level context* that is utilizable by context-aware services. This is achieved by mapping the acquired context information to an appropriate model and by generating new (or more) knowledge out of the model applying reasoning and filtering mechanisms or data fusion methods. A good overview about context modeling approaches is given by Strang and Linnhoff-Popien [146]. They list *key-value models*, *markup scheme models*, *graphical models*, *object-oriented models*, *logic-*

---

[3] http://maps.google.com/.

[4] http://www.android.com/.

based models as well as *ontology-based models* as possible approaches and come to the conclusion that ontologies are the most suitable way of modeling context. Based on this classification, Park and Kwon [112] compare the simplicity and representation power of various context models and argue that ontology-based context models "are superior [...] in showing hierarchies and logics [...] and is best in terms of model representation power". In a broad survey by Bettini et al. [22], not only modeling techniques are compared, but also reasoning approaches are considered. The authors claim that the limitations of the respective context modeling approaches can be reduced by integrating them to hybrid solutions. Examples are a *hybrid fact-based ontological model* or a *loosely coupled markup-based ontological model*. Section 2.5 focuses on a discussion of ontology-based context models as a foundation for our network and third-party context representations in Sects. 4.5.1 and 5.2.3.

An important aspect of context management is the quality of the processed context data, which has a direct impact when processing the acquired context information as well as on the user experience of the provided CAS. Buchholz et al. [29] introduce the notion of *Quality of Context (QoC)* and list *precision*, *probability of correctness*, *trustworthiness*, *resolution*, and *up-to-dateness* as possible and relevant QoC parameters. According to the authors, QoC enables a better CAS user experience by avoiding outdated or incorrectly sensed context data, for example. Furthermore, it supports the selection of appropriate context providers with high quality data and serves as an input for refining the context management process. Even though a variety of research has dealt with the QoC problem [99, 100, 108], Bellavista et al. [14] claim that the standardization of a QoC framework incorporating general as well as data-specific quality parameters is still an open issue.

Over the years, numerous papers have been published that present context management approaches as a whole within the mobile networks domain. Hochstatter et al. [67], for example, illustrate how context can be provisioned in mobile networks based on a conceptual framework. This framework separates context provisioning from service provisioning and enables "the rapid creation and deployment of new services". Floréen et al. [52], on the other hand, propose a *Context Management Framework* for mobile services that comprises separated functions for different tasks and reasoning methods. Influenced by this work, Moltchanov et al. [106] discuss context management requirements from the perspective of a mobile network operator and present a context representation formalism based on *ContextML* [79] as well as usage and application examples of their *Context Management Framework* at *Telecom Italia*. Mannweiler et al. [98] uses a context management architecture in order to utilize static as well as dynamic context information for enabling an efficient heteregeneous network access management. As presented by Schneider et al. [136], the same architecture also allows the integration of various context providers (e.g., sensors or mobile devices) in a "plug-and-play" manner for exploiting context information useful for network operators.

Based on in-depth studies and comparisons of a large amount of existing solutions in their survey, Bellavista et al. [14] propose a unified architectural model for context data distribution with *context data sources*, *context data sinks*, a *context management layer*, a *context data delivery layer* as well as a *runtime adaptation support* as its components. For enabling a better understanding, the authors also present a taxonomy for each layer. Perera et al. [115], on the other hand, highlight context management from the perspective of the *Internet of Things (IoT)*. In their extensive overview, the authors also analyze the vast majority of existing context management solutions (with the focus on IoT) and identify open issues such as the automatic connection of sensors to context management platforms, the automatic annotation of sensors, or the adoption of a "sensing-as-a-service model". Numerous other surveys [80, 95, 160] also deal with context management challenges from different perspectives.

In our work, we represent static as well as dynamic network context in the form of the *OpenMobileNetwork Ontology* (see Sect. 4.5.1), whereas the *Context Data Cloud Ontology* (see Sect. 5.2.3) comprises third-party context information. Furthermore, we apply several network context processing steps including also quality parameters for generating higher-level context out of the raw network measurements sensed from smartphones. For example, each low-level network context measurement that consist of a Cell-ID, a LAC, an MCC, and an MNC in combination with the current measurement position, is utilized in order to infer or update the position and coverage area of the corresponding mobile network cell.

2.2.2.3 Context Delivery and Services

Context Delivery comprises the process of making (higher-level) context information available for the usage within context-aware services. According to Hochstatter et al. [67], this can be achieved either in *pull* or *push mode*. In pull mode, the CAS actively requests context information from the platform in use whenever needed, whereas in push mode, context data is proactively delivered as soon as a predefined event occurs.

The interworking between context delivery modes and CASs can be perfectly seen in the area of location-based services [83], which belong to the most prominent class of CASs. LBSs are applications provided to users based on knowledge of their geographic location. Examples for such services include car navigation and traffic estimation, marketing, social networking, family and friend finder, travel and leisure activity recommendations, or games. There are also initiatives within the context of *Smart Cities* for enabling citizen participation in urban development through the use of LBSs [45].

In an LBS, the position of a target (e.g., a user) is determined by utilizing a number of existing positioning methods such as *GPS*, *Cell-ID* [37], or *WiFi Positioning* [38, 76]. These methods are applied in different combinations depending on the required accuracy of the application. The calculated location is usually represented in *World Geodetic System (WGS84)* coordinates (i.e., the latitude and longitude value on the globe) that are precise and easily processable by computer systems. A mapping between WGS84 coordinates and a human-readable form of a specific location (e.g., address, landmark, or place name) can be achieved through *geocoding* [56] and *reverse geocoding* services. In his book, Küpper [83] gives a very comprehensive overview about positioning methods and their mode of operation. Section 4.4, on the other hand, focuses on the scope of this thesis and provides a detailed analysis on different algorithmic approaches for estimating the position as well as coverage area of mobile network cells and discusses the advantages and drawbacks of each method.

According to Küpper et al. [84], LBSs can be functionally classified into different categories depending on their *user* or *service interaction* model, *user* and *target relationship*, *plurality*, *infrastructure*, and *environment* they are used in. The user or service interaction can either be *reactive* or *proactive*. A location-based service is reactive if information is delivered as soon as a user or service actively requests it, e.g., if a user asks for restaurant recommendations in the vicinity. In a proactive LBS, however, information is pushed to the service whenever certain conditions are met such as the automatic recommendation of a POI whenever the user enters a specific mobile network cell. The user or target relationship, on the other hand, is divided into *self-referencing* and *cross-referencing*. A service is called self-referencing when it relies on the position of the requesting user itself, whereas a cross-referencing LBS processes the location information of other targets (e.g., a friend tracking service). Basically, the differentiation is made on whether the LBS user and the tracked target is identical or different. Concerning the plurality, an LBS can make use of the location of a single user (*single-target*) or of many users (*multi-target*). Moreover, if the location information is collected and computed centrally, the LBS is said to be based on a *central infrastructure*. Alternatively, a service is defined as being *peer-to-peer* if it communicates directly with each participating node for exchanging location information. Finally, an LBS can be designed to work in outdoor or indoor environments.

Sathe et al. [130] define the next generation of *LBS 2.0* applications as services that utilize sensor-generated data of a variety of new mobile devices coming up (e.g., smart watches and wearables) in order to enhance the user experience and to provide information about the user's behavior and environment. Furthermore, they present possible challenges and opportunities in the future with a special focus on data management.

Over the last decade, the research community more and more understood the value of semantics for LBSs, so that many attempts were made to enrich LBSs with semantic information. For this purpose, *Semantic Web* standards (see Sect. 2.3) have been extensively used. In an early work, van Setten et al. [140], for instance, present the mobile tourist application *COMPASS* that is based on a platform using various ontologies describing context or providing a POI class hierarchy. Patkos et al. [113], on the other hand, propose a context-aware pedestrian guiding system based on a context model for representing persons, geographical spaces, location coordinates, and events. In a very recent study, Lee et al. [88] illustrate an LBS based on a *University Activity Ontology* that describes indoor activities in a university context. This service enables the calculation of the shortest path between indoor and outdoor spaces within a university campus and provides views in 2D as well as in 3D.

The importance of linking LBSs with semantics is outlined by Ilarri et al. [71] with a list of possible research areas. According to the authors, the implementation of an "intelligent query-answering approach that takes the user's context into account" or the definition of semantic locations as well as trajectories that abstract location information from geographic coordinates to higher-level concepts are two areas where usage of semantic technologies could be beneficial. The latter is a challenge that we addressed in [C9, C10] as well as in Sect. 7.2. Another work in this direction is presented by Toutain et al. [148], who focus on the formulation of a user location as well as the semantic context reasoning that can be performed on such a location description. Other areas that Ilarri et al. [71] list, comprise the "interoperability among LBSs and providers", location privacy, and "reasoning in complex and dynamic contexts".

Besides modeling ontologies, the applicability and added value of using *Linked Data* within LBSs is highlighted by the *DBpedia Mobile* app of Becker and Bizer [13]. This app displays *DBpedia* location data on a map based on the position of the user and allows to further navigate through interlinked information in other datasets. In addition, user-generated content can be uploaded and added to the *DBpedia* dataset. Another example for incorporating *Linked Data* into an LBS is given by Ostuni et al. [110]. They developed a context-aware movie recommendation system based on movie data from *DBpedia* [26] that also identifies theaters located nearby.

Even though there exist some work within this field, the combination of *Linked Data* and LBSs is still rather rare. In Sects. 7.2 and 8.2, we showcase several *Semantic Positioning* solutions that utilize the power of *Linked Data* and overcome the limitations of classic geocoding as well as geofencing methods and further add semantic features to proactive self-referencing and cross-referencing LBSs.

2.3 Semantic Web Technologies

The vision of the *Semantic Web* has its origins in a time when information on the Web was comprehendible for humans and only processable by computers in a syntactic form or from a structural perspective. The "meaning" as it was derivable for humans out of the context of the presented information, was not clear for machines due to the fact that a semantic enrichment of the data was missing.

In 2001, Tim Berners-Lee [21] described the *Semantic Web* as an extension of the WWW, "in which information is given well-defined meaning, better enabling computers and people to work in cooperation". It enables "computers to intelligently search, combine, and process Web content based on the meaning that this content has to humans" [66].

Many interpretations were made and different directions were taken throughout the time for realizing (aspects of) the *Semantic Web* [31, 101]. However, a major step towards the vision of a global and shared Web of machine-readable data (as originally intended by Berners-Lee [18]) has been taken by *Linked Data* [25]. Berners-Lee referred to this approach as the "Semantic Web done right" [19]. Later, the *World Wide Web Consortium (W3C)* renamed their vision of the *Semantic Web* into the *Web of Data* (see footnote 1 in Chap. 1).

Together with the vision, a stack of technologies have been standardized and recommended for the *Semantic Web*, such as RDF, *RDF Schema (RDFS)*, the *Web Ontology Language (OWL)*, SPARQL, and *Linked Data*, which are explained in the following subsections.

In this thesis, we share the same understanding of the *Semantic Web* as Berners-Lee and hence use the term "semantic" and its variations in the context of *Linked Data*. We refer to "semantically enriched" or "semantically modeled" data if it is represented using RDF, RDFS, or OWL and if it is interlinked with other datasets according to the principles of *Linked Data*. Furthermore, we consider the *LOD Cloud* as a huge pool of contextual information that can be leveraged for implementing context-aware services (see Sect. 2.3.5.1). In our case, a "semantically enriched context-aware service" exploits interlinked information in *Linked Data* format and adapts its behavior or the provided content based on the context of the user.

2.3.1 Resource Description Framework

The core technology of the *Semantic Web* is the *Resource Description Framework* [85, 121], which is a graph-based data model specifically designed to represent and interlink information on the Web in a flexible and easy-to-understand format.

2.3.1.1 Data Model

RDF is based on a triple pattern comprising a *subject*, a *predicate*, and an *object*, also known as an *RDF statement*. A collection of RDF statements is named *RDF graph*, while multiple RDF graphs form an *RDF dataset* that consists of exactly one unnamed *default graph* and a number of uniquely named RDF graphs.

Each part of a statement usually represents a *thing* that can be anything such as a real-world object, a document, an abstract concept, a number, or a string. It is synonymously also called a *resource* or an *entity*. A subject and object node stand for two resources that are related to each other, whereas the (binary) relationship between them is set using a predicate (property). Figure 2.4 depicts an exemplary RDF statement with a representation of a mobile network cell

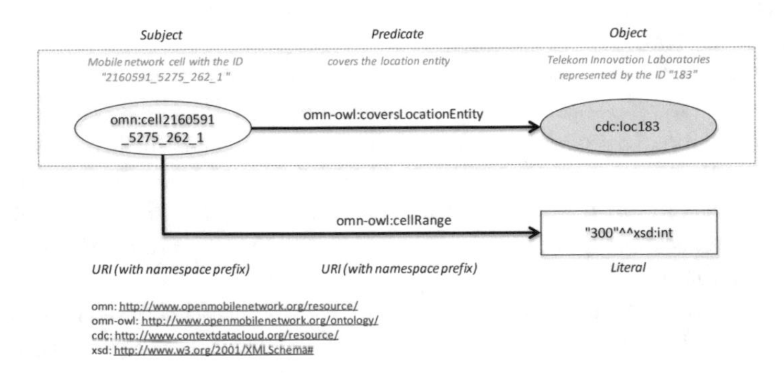

Fig. 2.4 Example for RDF statements

(omn:cell2160591_5275_262_1) as the subject, which is related to an identifier for a point of interest (cdc:loc183 as the object) using the property omn-owl:coversLocationEntity.

There are three types of identifiers that are utilized for representing the elements of a triple, namely *Internationalized Resource Identifiers (IRIs)* [47], *literals*, and *blank nodes*. Subjects are identified by using IRIs or blank nodes, whereas objects can also be additionally defined by literals. Predicates, on the other hand, are only representable via IRIs.

IRIs stand for a generalized form of *Uniform Resource Identifiers (URIs)* allowing the usage of the full Unicode string set (see Fig. 2.4 for several examples). They have the characteristic to be dereferenceable enabling a lookup of related information on a remote server. This concept is a very important aspect in Web architecture and hence of utter relevance when applying the principles of *Linked Data* (see Sect. 2.3.5).

A set of IRIs build an *RDF vocabulary* (also known as ontology) when specified to be used within an RDF graph for giving "meaning" to its resources. The entities within an RDF vocabulary usually share the first part of the whole IRI string, which is known as the *namespace IRI*. In order to improve readability, namespaces are also abbreviated by using *namespace prefixes*. As illustrated in Fig. 2.4, the exemplary prefix omn-owl represents a shortened form of http://www.openmobilenetwork.org/ontology/ that identifies the namespace of the *OpenMobileNetwork* vocabulary.

Literals, on the other hand, are values, such as strings, numbers, or dates, that consist of two main elements: the textual representation (i.e., the lexical form) in Unicode string and a *datatype IRI* associating a datatype to the literal value for making a correct processing of the data possible. Here, RDF mainly supports the datatypes specified in the *Extensible Markup Language (XML) Schema* [96]. An exemplary object with a literal is highlighted in Fig. 2.4 that maps the lexical form of 300 to an xsd:int datatype IRI standing for http://www.w3.org/2001/XMLSchema#int.

Blank nodes fulfill a similar function to variables in algebra and represent subjects or objects that denote resources without specifically identifying them with an IRI. They are often utilized as parent nodes for grouping data. However, blank nodes can only be used locally within an RDF dataset. Creating links to them from external RDF datasets is not possible, which is why their use is not recommended within the context of *Linked Data*.

2.3.1.2 Serialization Formats

In order to process an RDF graph and write the included triples down to a file, it needs to be serialized using (one of) the standardized RDF syntaxes. There are several well-known serialization formats, which are introduced in the next paragraphs and accompanied with examples using the RDF statements in Fig. 2.4.

The first standardized syntax is *RDF/XML* [54]. It is based on XML and structures RDF triples in an XML element starting with the rdf:RDF tag (see Listing 2.1 for an example). By using the xmlns:$PREFIX$ attribute within rdf:RDF, namespace prefixes (e.g., xmlns:omn-owl) are defined for a set of XML elements and attributes. An `rdf:Description` element, on the other hand, is utilized for specifying the subject of a triple accompanied by the `rdf:about` attribute standing for its IRI. Each sub-element within `rdf:Description` represents a triple with the corresponding predicates and objects for the given subject. If the object of a triple is identified by an IRI, the predicate sub-element (e.g., `omn-owl:coversLocationEntity`) has no content and the `rdf:resource` attribute is applied for specifying the object IRI. In case the object is a literal, the predicate sub-element is extended by an `rdf:datatype` attribute for mapping the datatype IRI of the object and the literal value (e.g., `300`) is used as the content of the predicate sub-element.

```
<?xml version="1.0" encoding="utf-8" ?>
<rdf:RDF
    xmlns:rdf="http://www.w3.org/1999/02/22-rdf-syntax-ns#"
    xmlns:omn-owl="http://www.openmobilenetwork.org/ontology/">
  <rdf:Description rdf:about="http://www.openmobilenetwork.org/
       resource/cell2160591_5275_262_1">
    <omn-owl:coversLocationEntity rdf:resource="http://www.
         contextdatacloud.org/resource/loc183" />
    <omn-owl:cellRange rdf:datatype="http://www.w3.org/2001/
         XMLSchema#int">300</omn-owl:cellRange>
  </rdf:Description>
</rdf:RDF>
```

Listing 2.1 RDF statements of Fig. 2.4 serialized in RDF/XML format

Even though the XML format provides a structured form for representing an RDF graph, it is not designed to be human-readable. Therefore, new serialization formats, such as the so-called *Turtle family of RDF languages*, came up focusing more on the readability aspect.

```
<http://www.openmobilenetwork.org/resource/cell2160591_5275_262_1>
  <http://www.openmobilenetwork.org/ontology/coversLocationEntity>
  <http://www.contextdatacloud.org/resource/loc183> .

<http://www.openmobilenetwork.org/resource/cell2160591_5275_262_1>
  <http://www.openmobilenetwork.org/ontology/cellRange>
  "300"^^<http://www.w3.org/2001/XMLSchema#int> .
```

Listing 2.2 RDF statements of Fig. 2.4 serialized in N-Triples format

The Turtle family of RDF languages supports four ways of serialization. *N-Triples* [138], for instance, provides a simple line-based syntax, where each line stands for a triple with a period at the end of the line marking a completed triple. Angle brackets (<>) comprise an IRI, whereas literals are denoted by attaching the literal value to the datatype IRI with a ^^ delimiter. Listing 2.2 illustrates the triples of Fig. 2.4 in N-Triples format.

N-Triples is extended by the *Terse RDF Triple Language (Turtle)* [33], which simplifies the serialization format in terms of writing and readability by supporting various syntactic shortcuts such as namespace prefixes, lists, and shorthands for datatyped literals. As exemplary shown in Listing 2.3, a *base IRI* (http://www.openmobilenetwork.org/resource/) is utilized in order to resolve a *relative*

IRI (cell2160591_5275_262_1) against it. In addition, namespace prefixes (e.g., omn-owl) are defined for IRIs, so that they can be abbreviated and used when writing down the triples (e.g., omn-owl:cellRange). Another shorthand is provided when constructing triples with the same subject (in our case cell2160591_5275_262_1). The subject is listed once and each predicate-object pair that follows it with a semicolon represents a triple using this subject. A period after a predicate-object pair marks the end of the triple set belonging to this subject.

```
BASE <http://www.openmobilenetwork.org/resource/>
PREFIX omn-owl: <http://www.openmobilenetwork.org/ontology/>
PREFIX cdc: <http://www.contextdatacloud.org/resource/>
PREFIX xsd: <http://www.w3.org/2001/XMLSchema#>

<cell2160591_5275_262_1>
  omn-owl:coversLocationEntity cdc:loc183 ;
  omn-owl:cellRange "300"^^xsd:int .
```

Listing 2.3 RDF statements of Fig. 2.4 serialized in Turtle format

TriG [33] and *N-Quads* [32] are two other serialization variants of the Turtle family of RDF languages. In contrast to Turtle that supports only single graphs, TriG enables the serialization of multiple named graphs forming an RDF dataset. For this purpose, the syntax of Turtle is extended by a GRAPH keyword including the graph name and curly braces that comprise the triples of the respective graph. N-Quads is similar to TriG in its function and provides support for multiple graphs by enriching the syntax of N-Triples with a fourth element for each triple. This element stands for the IRI of the named graph the respective triple is part of.

The serialization format named *RDFa* [64], on the other hand, provides means to embed RDF triples into *Hypertext Markup Language (HTML)* documents. This is useful in cases where publishers of RDF data are just able to modify HTML documents, but have restricted rights on the hosting infrastructure itself (e.g., managed content management systems) for configuring *303 redirects* and *content negotiation* as a way of publishing various descriptions for resources (see Sect. 2.3.5). RDFa extends HTML elements (e.g., div or span) by special attributes such as resource, property, typeof, or prefix. By doing so, it enables search engines to process the embedded RDF data and semantically enrich the search results.

JSON-LD [143] is one of the latest serialization formats standardized by W3C and enables documents in *JavaScript Object Notation (JSON)* [42] format to be mapped into RDF. As the name implies, it is specifically designed to be used in the context of *Linked Data*, which is why it provides a universal identifier mechanism that enables a JSON document to link an object of another JSON document on the Web. In addition, it supports the use of datatypes and language annotations for values.

2.3.2 RDF Schema

The RDF data model sets links between resources and thus enables the formulation of statements. However, it does not take into consideration for what these resources stand for in the real world. Adding semantic information to an RDF graph is achieved by associating its resources with RDF vocabularies that provide domain-specific concepts for representing (groups of) things in the real world and their relationships to each other.

RDF Schema [58] enables the creation of (lightweight) vocabularies for RDF by defining terms for classes and properties that describe groups and hierarchies of resources as well as their relationships. The concepts of RDFS are identified by the namespace http://www.w3.org/2000/01/rdf-schema# with the prefix `rdfs` as well as http://www.w3.org/1999/02/22-rdf-syntax-ns# abbreviated by `rdf`. This section gives a brief overview about the relevant concepts that are widely applied within the context of *Linked Data* and accompanies the descriptions by examples in Turtle format stated in Listing 2.4. Please note that the prefixes are omitted for readability purposes.

```
cdc:loc183
  rdf:type cdc-owl:LocationEntity .

omn-owl:coversLocationEntity
  rdfs:domain omn-owl:Cell ;
  rdfs:range cdc-owl:LocationEntity ;
  rdfs:subPropertyOf omn-owl:covers .
```

Listing 2.4 RDF statements in Turtle format expressed using RDFS

In RDFS, two major classes are defined, namely `rdfs:Class` and `rdf:Property`. An `rdfs:Class` represents all resources being *RDF classes*. A resource is defined as a class by setting a relation between the resource and the term `rdfs:Class` using the `rdf:type` predicate. By doing so, resources are structured into groups. Members of such a group (i.e., class) are also identified by using the `rdf:type` predicate and are known as instances of this class. Listing 2.4 exemplary shows the resource `cdc:loc183` being an instance of the class `cdc-owl:LocationEntity`.

RDFS also supports the description of class hierarchies (i.e., sub-class relationships). This is done by setting the `rdfs:subClassOf` property between two classes. By doing so, all instances of an exemplary class *B* that is an `rdfs:subClassOf` another class *A* (also known as the super-class in this context), become also instances of class *A*.

`rdf:Property`, on the other hand, stands for the class of all *RDF properties*. Similar to RDF classes, the predicate `rdf:type` provides the means to declare a property resource as an instance of `rdf:Property`.

Hierarchies for properties are described by the `rdfs:subPropertyOf` predicate. If two resources are related to each other using an exemplary property *D* (e.g., `omn-owl:coversLocationEntity` in Listing 2.4), which is the `rdfs:subPropertyOf` of another predicate *C* (e.g., `omn-owl:covers` in Listing 2.4), then the two resources are implicitly also related by the property *C*.

Two other relevant concepts of RDFS are `rdfs:domain` and `rdfs:range`. Declaring the `rdfs:domain` of a certain property implies that a subject resource attached to this property is always an instance of one or more specific classes, whereas the usage of the `rdfs:range` for a certain property states that the object resource is always a member of one or more specific classes. In Listing 2.4, for example, we can see that the `rdfs:domain` of the `omn-owl:coversLocationEntity` property is `omn-owl:Cell` and the `rdfs:range` is defined as `cdc-owl:LocationEntity`. This means that the subject of a triple comprising `omn-owl:coversLocationEntity` as its predicate always belongs to the class of `omn-owl:Cell`, whereas the object is an instance of `cdc-owl:LocationEntity`.

By using RDFS, structure and semantic information is added to an RDF graph. It enables additional information to be implicitly queried without having all triples explicitly created within the graph. This supports simplicity in RDF graphs by keeping its full comprehensiveness at the same time.

2.3.3 Web Ontology Language

The expressiveness of RDF Schema is extended by the *Web Ontology Language* [137], which uses http://www.w3.org/2002/07/owl# as its namespace with the prefix `owl`. It adds more concepts for describing classes and properties such as disjointness, cardinality, equality, richer typing and characteristics of properties, and enumerated classes. Please note that this section briefly highlights specific aspects of the first version of OWL that are relevant for this thesis. A comprehensive overview of *OWL 1* is given in [137].

OWL comprises three sub-languages that differ in the depth of the provided expressiveness and formal complexity. *OWL Lite*, for example, is used for knowledge domains where a classification hierarchy and simple constraints are sufficient. *OWL DL*, on the other hand, keeps the balance between high expressiveness and computational completeness by allowing the use of all provided concepts under certain restrictions. For instance, while a class can be a sub-class of another class, it cannot be an instance of a class. In *OWL Full*, maximum expressiveness is supported. However, no guarantees are given that all conclusions are computable.

Linked datasets are usually represented using the expresiveness of OWL Lite since they rather exploit ontologies as published lightweight schemas in contrast to the initial understanding of the *Semantic Web*, where ontologies were considered to be highly expressive models. This shift in the usage mindset mainly arose after realizing that semantic technologies (as originally intended to be used) are not mature enough to be very beneficial in productive environments [43]. Therefore, we focus on the vocabulary terms of OWL Lite and do not go into OWL DL as well as OWL Full.

In addition to the class and property hierarchy of RDFS, OWL defines terms that describe equality and inequality of classes as well as individuals. `owl:equivalentClass` and `owl:equivalentProperty`, for example, are applied for representing synonymous classes and properties having the same meaning. With `owl:equivalentClass`, instances of a class *A*, which is equivalent to a class *B*, are defined as being also instances of class *B* and vice versa.

A similar term on instance level is `owl:sameAs` stating that two individuals with different textual representations are the same. This concept is of utter importance in the context of *Linked Data* since a huge number of links between datasets in the *LOD Cloud* are created using `owl:sameAs` in order to describe that a resource in dataset *A* and another resource in dataset *B* have the same meaning [59]. Frameworks for automatically generating links between linked datasets [151] usually work with `owl:sameAs` by default.

If it is of importance to state that two individuals (of the same class, for example) are different from each other, the concept `owl:differentFrom` is used.

Another added value of OWL is the specification of property characteristics. Here, OWL fundamentally distinguishes between `owl:DatatypePropertys` and `owl:ObjectPropertys`. An `owl:DatatypeProperty` stands for a predicate that connects an individual to a datatyped literal (see Sect. 2.3.1), whereas an `owl:ObjectProperty` creates a relation between two individuals. Other characteristics describe that two properties are inverse to each other (e.g., `omn-owl:operatedBy owl:inverseOf omn-owl:operatesCell`), are transitive (`owl:TransitiveProperty`), or symmetric (`owl:SymmetricProperty`), for example.

Concepts for property restrictions (`owl:allValuesFrom` or `owl:someValuesFrom`) and cardinality features (e.g., `owl:minCardinality` or `owl:maxCardinality`) are also defined by OWL. However, these terms are not explained further in detail since they are not used within this thesis.

In 2012, the *W3C OWL Working Group* specified *OWL 2* [152] adding more features by preserving backwards compatibility at the same time.

2.3.4 SPARQL

SPARQL [139] stands for *SPARQL Protocol And RDF Query Language*. As the name implies, it provides a graph-based query language for retrieving RDF data.

```
PREFIX omn-owl: <http://www.openmobilenetwork.org/ontology/>
PREFIX omn: <http://www.openmobilenetwork.org/resource/>
PREFIX cdc: <http://www.contextdatacloud.org/resource/>
PREFIX rdfs: <http://www.w3.org/2000/01/rdf-schema#>
PREFIX dbpedia: <http://dbpedia.org/resource/>

SELECT ?cell ?technology ?cellrange
WHERE {
  ?cell rdf:type omn-owl:Cell .
  ?cell omn-owl:cellType ?technology .
  ?cell omn-owl:cellRange ?cellrange .
  ?cell omn-owl:operatedBy ?mnc .
  ?mnc rdfs:seeAlso dbpedia:T-Mobile .
  ?cell omn-owl:coversLocationEntity cdc:loc183 .

  FILTER (?cellrange > 1000)
}
ORDER BY ASC(?cellrange)
```

Listing 2.5 SPARQL example: Querying *sparql.openmobilenetwork.org* for all *Telekom* mobile network cells with a coverage area radius greater than 1000 m that cover *T-Labs* at Ernst-Reuter-Platz in Berlin

A SPARQL query is based on a set of triple patterns similar to RDF - also called a basic graph pattern - where each subject, predicate, and object can be substituted by a variable in the form `?variable`. It consists of two building blocks: The first part comprises the `SELECT` clause including all variables that are ultimately listed in the query result. Alternatively, SPARQL also allows the usage of other query forms such as `CONSTRUCT`, `ASK`, or `DESCRIBE`. In the second building block, the `WHERE` clause is used in order to map the basic graph pattern to the RDF data graph. This part also incorporates `FILTER` functions enabling the restriction of the result set by using regular (`regex`) or arithmetic expressions. In addition, functions for grouping (`GROUP BY`) and ordering (`ORDER BY`) the result set, or setting limits (`LIMIT`) and offsets (`OFFSET`) are provided.

The *OpenMobileNetwork* dataset query example in Listing 2.5 showcases a typical SPARQL query including the two building blocks and commonly used functions. This query requests all *Telekom* mobile network cells with a radius for a coverage area greater than 1000 m that cover *Telekom Innovation Laboratories* at Ernst-Reuter-Platz in Berlin, Germany.

1. Prefixes for the namespaces are defined in the form `PREFIX omn-owl: <http://www.openmobilenetwork.org/ontology/>` allowing the usage of shorthands within the triple patterns.
2. The `SELECT` clause comprises the variables `?cell`, `?technology`, and `?cellrange` that ultimately form the query result.
3. The `WHERE` clause is used in order to map the basic graph pattern to the *OpenMobileNetwork* data graph.

   - The first triple pattern represents all instances (`?cell`) of the class `omn-owl:Cell` using the `rdf:type` predicate.
   - The second pattern looks for all object individuals (`?technology`) that are related to the `?cell` instances with the property `omn-owl:cellType`.
   - All objects (`?cellrange`) that are linked to the `?cell` instances with the property `omn-owl:cellRange` are identified in the third triple pattern.

Table 2.1 Excerpt of the results for the SPARQL query in Listing 2.5

cell	technology	cellrange
omn:cell55959_5126_262_1	omn:EDGE	1032
omn:cell44932_5126_262_1	omn:EDGE	1074
omn:cell2159548_5275_262_1	omn:UMTS	1077
omn:cell2117699_5275_262_1	omn:UMTS	1104

- Similar to the ones above, the fourth triple pattern represents all object resources (`?mnc`) that are related to the `?cell` instances using the `omn-owl:operatedBy` predicate.
- The fifth triple pattern narrows the `?mnc` entities to only those that are linked to the `dbpedia:T-Mobile` resource by `rdfs:seeAlso`.
- Finally, the last pattern represents only the `?cell` instances that are connected to the `cdc:loc183` resource via `omn-owl:coversLocationEntity`.
- The `FILTER` function numerically restricts the `?cellrange` literals. Only those literals are considered that have a value greater than `1000`.
- The `ORDER BY` function enables a sorting of the results (`?cellrange`) in ascending order.

An excerpt of the result set is depicted in Table 2.1. As we can see, the result set consists of three columns that are named according to the variables in the `SELECT` clause. Only *Telekom* cells (identified by `262_1`) are listed in the `cell` column with a coverage area radius greater than 1000m. The mobile network technology for each cell is given in the `technology` column. In addition, the radius of the coverage areas are sorted in ascending order in the `cellrange` column. Please note that due to size issues within this manuscript, the results are illustrated using the prefixes. Normally, SPARQL uses the whole namespace when listing the query results.

SPARQL provides much more (sophisticated) features. For a comprehensive overview, please refer to [139].

2.3.5 Linked Data

Tim Berners-Lee coined the term *Linked Data* in 2006 as his vision of the "Semantic Web done right" [19] describing an approach for publishing and interlinking structured data on the Web based on the key standards of the WWW [153] (e.g., URIs, the *Hypertext Transfer Protocol (HTTP)*, HTML, and hypertext links). He introduces this approach in one of his personal notes [20] and summarizes the principles of *Linked Data* by giving four basic rules:

1. Use URIs for naming things.
2. Use HTTP URIs for enabling people to look up those names.

3. Provide useful information using RDF and SPARQL when people look up those URIs.
4. Link things to other URIs, so that people can discover related things.

As a key literature in this field, Heath and Bizer detail the notes of Berners-Lee and present the concepts as well as technologies behind *Linked Data* in [62], which we outline in the following paragraphs.

Referring to the first two rules, *things* are real-world objects that are called *resources* in the context of *Linked Data*. Each resource is identified by an HTTP URI that enables a straightforward approach of creating unique names and at the same time allows the access to further information related to the resource by using the HTTP protocol.

In contrast to the *Web of Documents*, *Linked Data* is not only intended to be used by humans, but also by machines. Therefore, several representations (descriptions) of the resource are offered when dereferencing (i.e., looking up) its URI. For humans, HTML is used as a representation for providing information to the resource, while machines rather process RDF data describing the same entity. This is done by utilizing the *content negotiation* mechanism as standardized in [51] that enables a server to identify via the HTTP header what kind of document is preferred by the requesting HTTP client.

There are two ways of defining HTTP URIs, namely *303 URIs* and *Hash URIs* [131]. Due to the fact that a URI of a real-world object is not directly dereferenceable, the 303 URI method uses a *303 redirect* and - as the name implies - redirects the client's initial request to a preferred URI (selected based on content negotiation) describing the resource in a human- or machine-readable format.

```
<?xml version="1.0" encoding="utf-8" ?>
<rdf:RDF
    xmlns:rdf="http://www.w3.org/1999/02/22-rdf-syntax-ns#"
    xmlns:rdfs="http://www.w3.org/2000/01/rdf-schema#"
    xmlns:omn-owl="http://www.openmobilenetwork.org/ontology/"
    xmlns:ogc="http://www.opengis.net/ont/geosparql#"
    xmlns:geo="http://www.w3.org/2003/01/geo/wgs84_pos#" >
  <rdf:Description rdf:about="http://www.openmobilenetwork.org/resource/wifiap376138223">
    <omn-owl:isCoveredBy rdf:resource="http://www.openmobilenetwork.org/resource/cell
           2160591_5275_262_1" />
  </rdf:Description>
...
```

Listing 2.6 Parts of the RDF/XML document delivered after a 303 redirect for http://www.openmobilenetwork.org/resource/cell2160591_5275_262_1

Taking a resource of the *OpenMobileNetwork* (see Sects. 4.5.1 and 6.3.3.1) as a practical example, if a *Linked Data* client tries to dereference http://www.openmobilenetwork.org/resource/cell2160591_5275_262_1, which is a URI for identifying a mobile network cell, the server sends the HTTP response status code `303 See Other` along with http://www.openmobilenetwork.org/data/cell2160591_5275_262_1 representing an RDF/XML description of the resource. Dereferencing this URI in the second step delivers the RDF/XML document as exemplary shown in Listing 2.6.

Hash URIs, on the other hand, address the main drawback of 303 URIs triggering two HTTP requests in order retrieve the description of a single real-world object.

This is done by adding special *fragment identifiers* to the URI using a hash symbol (#). Here, each fragment identifier represents a resource or an ontology concept, such as http://www.w3.org/2003/01/geo/wgs84_pos#SpatialThing, identifying the vocabulary term for "anything with spatial extent" within the *WGS84 Geo Positioning Vocabulary*[5] [27]. When sending an HTTP request, the fragment part (e.g., `#SpatialThing`) of the URI is removed leading to a URI (e.g., http://www.w3.org/2003/01/geo/wgs84_pos) that delivers an RDF or HTML document (selected based on content negotiation) of all concepts and resources sharing the same basic part.

By using Hash URIs, ambiguity is avoided when providing several representations of a real-world object since one URI is used to identify the real-world object as well as the human- and machine-readable descriptions. However, if the ontology or the dataset exceeds a certain amount of triples, large amount of (unnecessary) data is retrieved everytime a certain resource is requested. Therefore, Hash URIs should be defined for (small) vocabularies, whereas 303 URIs should be used when a dataset or ontology reaches a significant size.

Providing useful information when dereferencing URIs is the third principle of *Linked Data*. This is accomplished by utilizing the RDF data model including its various serialization formats (as illustrated in Sect. 2.3.1) as well as SPARQL as a query language (see Sect. 2.3.4) for retrieving semantic data. One advantage of using RDF is that its graph model based on Web standards as well as its interlinking nature is "inherently designed for being used at global scale", where each RDF triple plays a role in the global *Web of Data* by becoming a potential entry point to more related data. Furthermore, it allows the usage of different schemata for giving concepts as well as resources more meaning and structuring the data according to the needs of the application domain.

The most interesting and innovative part of the *Linked Data* paradigm is outlined within the forth principle given by Berners-Lee. Connecting heterogeneous datasets by interlinking resources of different origin to each other extends the Web to a single "global data space". An external RDF link is set by relating the subject of a triple in the namespace of one dataset to an object of a triple in the namespace of another. The predicate in between could also have a different namespace. This approach of dereferencing URIs with other namespaces enables an exploration of the *Web of Data* in a *follow-your-nose* fashion.

Different types of RDF links can be set between resources and concepts as exemplary illustrated in Fig. 2.5. A *relationship link* connects related information to a real-world object such as a certain POI in the dataset of *Linked Crowdsourced Data* (`cdc:loc183`) being covered (`omn-owl:coversLocationEntity`) by a specific mobile network cell (`omn:cell2160591_5275_262_1`) in the *OpenMobileNetwork* dataset. *Identity links*, on the other hand, identify the same real-world object in different datasets. Using the predicate `owl:sameAs`, the OMN resource `omn:HSPA`, for example, is linked to `dbpedia:Evolved_HSPA` in *DBpedia* being a different representation of the same resource. The last type of

[5] *geo*, http://www.w3.org/2003/01/geo/wgs84_pos#.

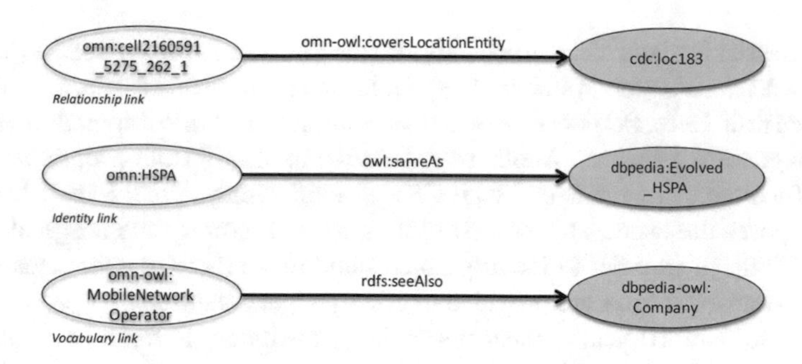

Fig. 2.5 Examples for external RDF links

vocabulary links uses relations between vocabulary terms for providing more meaning to ontology concepts and reusing existing terms in order to avoid redundancy in vocabulary definitions. For this purpose, the example in Fig. 2.5 interlinks the concept `omn-owl:MobileNetworkOperator` to the term `dbpedia-owl:Company` with an `rdfs:seeAlso` property defining a network operator also as a company described in *DBpedia*. This information is not directly required for our application areas, so that we do not see the necessity of describing a network operator in more detail within our dataset. However, having this link enables others to look up more information about an `omn-owl:MobileNetworkOperator` if interested.

In contrast to other data sources (e.g., Web APIs) and data structuring formats (e.g., *Comma-separated Values (CSV)*) where consumers of the data are forced to implement against proprietary APIs or parse different data formats to the desired model, *Linked Data* relies on a unified data model based on RDF that is designed to be comprehensible to humans as well as machines and to be used at global scale. Using Web standards like HTTP, URIs, and hyperlinks, it further enables a standardized approach for accessing data and allows the interlinkage as well as discovery of heterogeneous datasets.

2.3.5.1 Linking Open Data Cloud

As described above, *Linked Data* uses RDF for publishing data in the form of *subject-predicate-object* triples and for interlinking resources with entities from other datasets. By doing so, the Web is extended to a single "global data space" that is also known as the *Web of Data* [62].

However, in order to establish a global *Web of Data*, which is consumable by everybody similar to the *Web of Documents*, it is important that available data around the world is published with an open license according to the principles illustrated in Sect. 2.3.5. For this purpose, Berners-Lee encourages data owners by proposing a *5-star rating scheme* [20] describing the evolution of non-structured data to becoming *Linked (Open) Data*.

The first star is given if data is available on the Web with an open license despite its format, whereas the second star is added if this data is published as machine-readable structured data. Here, Berners-Lee gives the example of an "Excel spreadsheet instead of an image scan of a table". Another star comes into play if a non-proprietary format is used for structuring the data, i.e., if CSV is applied rather than Excel. The fourth star comprises the aspects of the first three stars and requires the usage of RDF as well as SPARQL in order to identify objects and make them dereferenceable. Five stars are awarded if links are provided to other people's datasets.

For more than 10 years, many researchers, institutes, libraries, etc. published their data according to the *Linked Data* principles covering various content domains, such as *Media*, *Life Sciences*, *Geography*, or *Government*, all directly linked to each other building a huge graph of interlinked structured data, which is also known as the *Linking Open Data Cloud* [2]. The number of datasets in this cloud has been increasing rapidly and has reached a total of 1,139 by February 2017. Figure 2.6 shows the latest update of *LOD Cloud* diagram.

These datasets provide a huge pool of contextual information that can be exploited by mobile network operators for correlating mobile network information with heterogeneous data sources and thus for providing semantically enriched context-aware services.

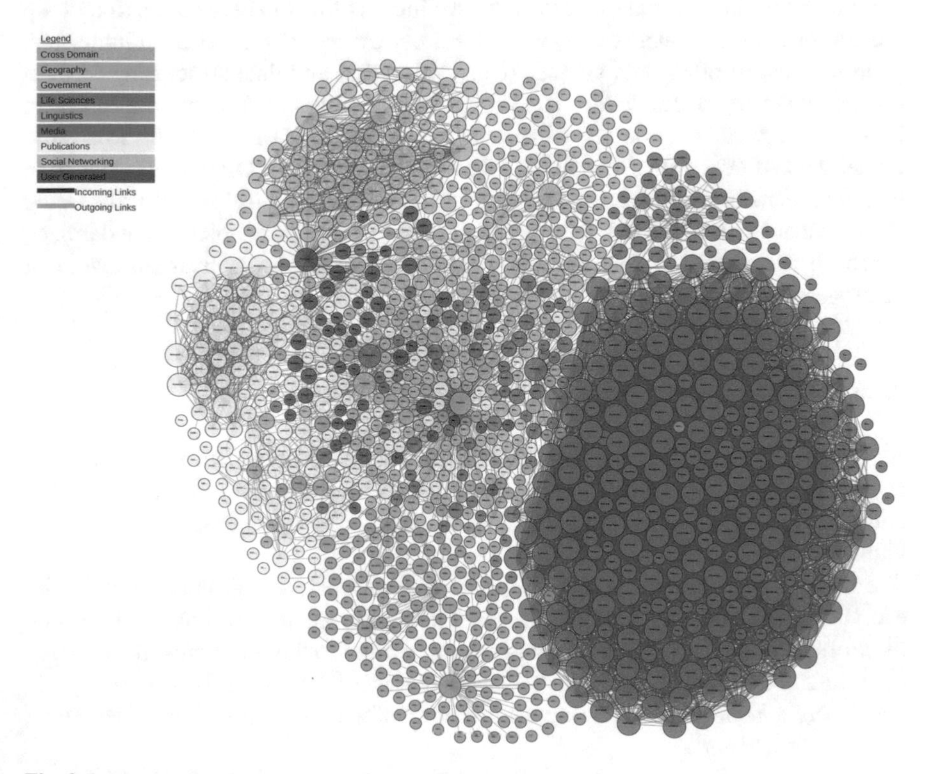

Fig. 2.6 Linking Open Data Cloud diagram, February 2017 [2]

2.4 Related Platforms and Datasets

We studied existing work in the areas of cellular and geo-related datasets by analyzing both external data sources (e.g., Web APIs or database dumps) as well as datasets within the *LOD Cloud*, where we focused on finding content similar or related to the datasets of the *OpenMobileNetwork* and *Linked Crowdsourced Data*. This kind of information is mainly available within the *Geography* domain, which counted 64 datasets at *datahub*[6] by April 14th, 2017. Other datasets that are also applicable as third-party context are not in the focus of this section.

Our analysis began in Sect. 4.1 with a detailed overview about available network data sources as a foundation for our conceptual work. Here, we described valuable network information that mobile network operators already possess and evaluated several cell as well as WiFi databases that are (partly) publicly available on the Web. In a detailed discussion, we identified drawbacks of these datasets such as access restrictions, unavailable network context parameters, or missing quality evaluations.

Thinking of smartphones as devices with sensors delivering (network) data (see Sect. 4.3), we also had a deep look into the *Geography* domain of the *LOD Cloud* and discovered several datasets (and platforms behind them) publishing various types of semantically enriched sensor data for enabling a variety of weather warning and natural disaster management use cases.

As an initial step towards the vision of the *Web of Things*[7], Le-Phuoc et al. [86] introduce an approach for publishing sensor data according to the principles of *Linked Data* enabling the creation of sophisticated mashups and continue their work by presenting a *Live Linked Open Sensor Database* [87] where the meaning of the collected sensor data is described by adding metadata and by interlinking it with other resources of the *LOD Cloud*.

Patni et al. [114], on the other hand, propose *Linked Sensor Data*[8] as an initiative publishing sensor descriptions and observations of weather stations in *Linked Data* format. In addition, Barnaghi and Presser [11] present *Sense2Web* as an alternative platform for describing sensors in RDF and interlinking this data to other data sources, while Yu and Liu [159] illustrate a solution for linked geo-sensor data within the water resources application domain based on the *W3C Semantic Sensor Network Ontology*[9] [41, 147]. Furthermore, a recent survey about approaches for linked streaming data [93] lists several other activities for generating (real-time and non real-time) linked sensor data and discuss lessons learned as well as open challenges.

None of these datasets in the *LOD Cloud* (and the systems behind them) consider publishing semantically enriched mobile network and WiFi access point topology data incorporating static as well as dynamic network context parameters (e.g., traffic or number of users in a radio cell). This makes the *OpenMobileNetwork* very unique in its form since it is the first (and only) dataset in the *LOD Cloud* comprising this kind

---

[6] https://datahub.io/dataset?tags=geographic&q=lod&tags=lod.

[7] Web of Things at W3C, https://www.w3.org/WoT/.

[8] http://wiki.knoesis.org/index.php/SSW_Datasets.

[9] *ssn*, http://www.w3.org/ns/ssn/.

of data that also tackles the challenges of the available cell and WiFi databases mentioned in Sect. 4.1 (e.g., access restrictions, unavailable network context parameters, or missing quality evaluations).

Switching to data sources related to *Linked Crowdsourced Data*, there are several commercial providers having datasets that include real user location preference data. *Foursquare* is the pioneer in this area that started to collect location check-ins of users through their smartphone apps back in 2009. The user community is being motivated with gamification methods by collecting badges or competing for mayorships. By doing so, *Foursquare* achieved to collect more than 10 billion check-ins from over 50 million users.[10] Later, *Facebook* also launched their *Places*[11] service as an integral part of their social network platform allowing users to check-in to places. *Yelp* is another service in this area more focusing on collecting reviews and ratings for places with an average of 89 million monthly unique visitors in Q4 2016[12], whereas *Google* provides location data through *Google Places*.[13]

All of these datasets offer high quality location (and check-in) data, which can also be requested through their APIs. However, a deeper look at their terms of use (see footnote 10 and 11 in Chap. 1)[14] reveals that storing this data in another database, combining or modifying it, is not allowed. In addition, these datasets are neither available in RDF format nor published as *Linked Data* restricting their use within the *LOD Cloud*.

As a counterpart to the commercial providers, *OpenStreetMap (OSM)*[15] is an open data project providing a huge amount of geographical data that is collected and edited by a crowdsourcing community. It is distributed under the *Open Data Commons Open Database License (ODbL)*[16] allowing free usage of the data as long as OSM is mentioned as the source of information. The data quality and the richness of detail differs from country to country and strongly depends on the number of actively contributing users. Nevertheless, for a vast amount of countries, the quality is comparable to commercial providers making OSM a valuable alternative for open data projects. Therefore, we have chosen OSM as the basis for our dataset and extended it with real user location preference data.

Within the *Geography* domain of the *LOD Cloud*, *LinkedGeoData* [144] and *GeoNames* [155] are the most prominent examples for geo-related datasets. *LinkedGeoData* publishes the data within OSM according to the principles of *Linked Data*. The dataset comprises more than 3 billion nodes, 300 million ways, and 20 billion RDF triples. M. Wick, on the other hand, presents *GeoNames*[17] as a "Gazetteer aggre-

---

[10]https://www.foursquare.com/about.

[11]https://www.facebook.com/places/.

[12]http://www.yelp.com/about.

[13]https://developers.google.com/places/.

[14]https://www.yelp.com/developers/api_terms.

[15]http://www.openstreetmap.org/.

[16]http://opendatacommons.org/licenses/odbl/.

[17]http://sws.geonames.org/.

gator of open geo data" [155] consisting of over 10 million geographical names[18] that can be requested through a set of Web services or downloaded in the form of database dumps including RDF.

In a more recent work, Gazzè et al. introduce *Tourpedia*[19] [55] as a "Wikipedia of Tourism", which incorporates RDF information about tourism places in eight European cities extracted from social media. In addition to static location information, *Tourpedia* calculates the sentiment for a place out of the social media reviews and attaches this information as a polarity value, which is a novelty in contrast to other datasets. However, due to the licensing problem mentioned above, data going beyond standard location information (e.g., reviews) is not published within the dataset, but rather links are provided to the social media origin.

Whether being big or small, all of these geo-related datasets within the *LOD Cloud* have in common that they do not connect dynamic parameters (e.g., check-ins, rating, or comments) as well as specific context situations (e.g., weather conditions, holiday information, or measured networks) to the static location data making *Linked Crowdsourced Data* an added value not only for mobile network operators, but also for the LOD community in general.

2.5 Related Context Ontologies

In general, ontologies enable the creation of a common understanding and representation of knowledge domains by using a set of class, property, and relation constructs. They are particularly suitable for modeling, structuring, and storing context information for diverse application domains as they provide the means for a shareable understanding and conceptualization of context.

Numerous context ontologies have been created so far; some more application-specific and others with the claim of being generic. In the following, we present related work in the areas of generic-purpose context ontologies (see Sect. 2.5.1), geo ontologies (see Sect. 2.5.2), mobile ontologies (see Sect. 2.5.3) as well as user profile and preference models (see Sect. 2.5.4). Furthermore, we briefly put the added value of the *OpenMobileNetwork Ontology* and the *Context Data Cloud Ontology* (as designed in Sects. 4.5.1 and 5.2.3) into perspective and highlight the influence of the related work to our contribution. Please note that a clear separation of the presented ontology models cannot be made since all of them combine various aspects of context, so that a lot of concepts within the ontologies are repeating themselves. Nevertheless, we try to categorize them according to the main purpose they serve.

[18]http://www.geonames.org/about.html.

[19]http://www.tour-pedia.org/.

2.5.1 Generic Context Ontologies

As mentioned above, countless ontologies have been developed throughout the years representing different facets of context in ubiquitous environments. Examples are the *COBRA-ONT* [35], *SOUPA* [36], *CONON* [154], and *CACOnt* [161] ontologies. By defining a set of context-specific ontologies and/or different types of ontology layers, these vocabularies either separate context domains (e.g., user, location, or activity context) from each other or generic context concepts from rather application-specific aspects. Chen et al. [35], for instance, follow the first approach with their *COBRA-ONT Ontology* and define a collection of application-specific ontologies for places as well as agents including their location and activity context.

The *CONON Ontology* [154], on the other hand, is an example for the latter approach comprising an upper ontology for generic context facets that is extended by specific ontologies to be used within different application domains. A similar approach is presented by Chen et al. [36] separating the *SOUPA Ontology* into a *SOUPA Core* and *SOUPA Extension* as well as by Xu et al. [161] with the *CACOnt Ontology* including a *user*, *device*, *service*, *space*, and *environment model*. Rodríguez et al. [125] also propose a multi-dimensional ontology model consisting of a *user context*, *Web services*, and *application domain ontology*. The user context ontology summarizes information about the user's device, position, occupation as well as service interests.

Dynamic aspects of context are in the scope of Riboni and Bettini [124]. The authors introduce the *ActivO Ontology* as part of the *COSAR* system that enables the recognition of recurring user activities (based on a use case in the health domain). Similar to that work, Scerri et al. [7, 133] model *live context* and *recurring personal situations* with their *DCON Ontology* for situation recognition purposes. This ontology describes *low-level*, *mid-level* as well as *high-level context* and reuses several vocabularies for representing personal context information.

Besides our work, Yu et al. [158] apply the principles of *Linked Data* for modeling context and explicitly treat the *LOD Cloud* as a context provider for enriching information within their model with available *LOD Cloud* context data. They present a *Linked Context* model that combines *user context* with *service context* by adopting and orchestrating a number of existing models as well as ontologies. Service provisioning is facilitated by annotating context data with information available in the *LOD Cloud*.

For the design of the *OpenMobileNetwork Ontology* and the *Context Data Cloud Ontology*, we follow the best practices of the related work and separate our ontologies into different context facets enabling a flexibility when using the data for context-aware services. Furthermore, in contrast to other ontologies that do not (fully) apply the principles of *Linked Data*, we utilize the capabilities of this approach by publishing our ontologies (as well as the instance data) and interlinking it with valuable information within the *LOD Cloud* for realizing a variety of innovative context-aware services.

2.5.2 Geo Ontologies

For representing mobile as well as WiFi networks from a topological perspective (as it is done in the *OpenMobileNetwork Ontology*) and for describing location-related context (as part of the *Context Data Cloud Ontology*), a geographic representation of the modeled data is essential.

A lot of research has been done in representing geospatial concepts with semantic technologies. In 2003, Brickley [27] specified a lightweight RDF geo vocabulary as part of the *W3C Semantic Web Interest Group (SWIG)*. The *WGS84 Geo Positioning Vocabulary* (see footnote 5 in this chapter) comprises predicates for latitude (`geo:lat`), longitude (`geo:long`) as well as altitude information in WGS84 format and is purposely kept very simple for basic geographic representations. The concepts of this ontology are widely used in linked datasets as well as in other geo-related vocabularies such as the *GeoNames Ontology*[20]. Nevertheless, in order to answer the need for sophisticated geospatial concepts, Lieberman et al. [92] extended this basic schema by defining the *W3C Geospatial Vocabulary*[21]. This ontology maps the features of the *Geographically Encoded Objects for RSS feeds (GeoRSS)* to OWL and hence provides concepts for `georss:points`, `georss:lines`, or `georss:polygons`, for example. However, support for terms that describe spatial relationships, such as *equals*, *disjoint*, *intersects*, or *contains*, is not given.

The next step towards a common representation of geospatial data is taken by Salas et al. [129]. They proposed the *NeoGeo Geometry Ontology*[22] and *NeoGeo Spatial Ontology*[23] enabling the usage of not only geometric shapes, but also spatial relationships.

The *Open Geospatial Consortium (OGC)*[24], on the other hand, did not only focus on providing an ontology, but rather worked on a comprehensive standard for "representing and querying geospatial data on the *Semantic Web*". *GeoSPARQL* [116] consists of a vocabulary[25] as well as a set of extensions for SPARQL. The *OGC GeoSPARQL Vocabulary* defines `ogc:SpatialObject` as a super-class with `ogc:Feature` and `ogc:Geometry` as its sub-classes. An `ogc:Feature` is a concept for describing a real-world object that has a geometric representation, such as a university complex. The representation is achieved by linking one or several `ogc:Geometrys` to the `ogc:Feature` depending on the complexity of its geometry (e.g., point or polygon). Geometry literals are defined either in *Well-known Text (WKT)* [94] or *Geography Markup Language (GML)* [118] format. For this purpose, the `ogc:asWKT` and `ogc:asGML` properties are used with `ogc:wktLiteral` and `ogc:gmlLiteral` as the datatypes for the literal values.

---

[20] *gn*, http://www.geonames.org/ontology/.

[21] *georss*, http://www.georss.org/georss/geo_2007.owl.

[22] *geom*, http://geovocab.org/geometry#.

[23] *spatial*, http://geovocab.org/spatial#.

[24] http://www.opengeospatial.org/.

[25] *ogc*, http://www.opengis.net/ont/geosparql#.

Please note that more or less, all of these vocabularies claim the `geo` namespace prefix for themselves. Therefore, in order to avoid confusion, we did not use the recommended namespace prefixes within the discussion above as well as in the design of our ontologies.

Several concepts of most of these ontologies are reused within the *OpenMobileNetwork Ontology* as well as the *Context Data Cloud Ontology*. Due to simplicity reasons and the fact that it sufficed our requirements, the latitude and longitude terms of the *WGS84 Geo Positioning Vocabulary* are utilized for describing the position of mobile network cells, WiFi access points, and POIs, while the polygonal coverage areas of the mobile network cells are mapped onto concepts of the *OGC GeoSPARQL Vocabulary*. The representation of a boundary box as defined by the *W3C Geospatial Vocabulary*, is part of the *Context Meta Ontology* that is introduced in Sect. 12.3.1.

Here, one could ask whether such a heterogeneity in reusing vocabulary concepts is necessary for describing similar geo-related aspects or if it completely sufficed to use one geo vocabulary that captures all our design requirements. The answer is that it may have sufficed to stick with one ontology. However, the development of our ontologies happened in a longer period of time and requirements changed during this time frame. Furthermore, in some cases, the specification and textual representation of a concept from one ontology fitted more to our design than the one of another schema. This also led to the fact that many ontologies have been reused for describing contextually similar aspects.

2.5.3 Mobile Ontologies

Before modeling the *OpenMobileNetwork Ontology*, we evaluated several ontologies and models that have been created to describe aspects of the mobile and telecommunications domain. A variety of them focus on different characteristics and capabilities of mobile devices. They typically comprise information about the supported communication standards (e.g., UMTS or Bluetooth) or device-specific characteristics such as the resolution of the display, the operating system, the battery capacity, or processing power.

Bandara et al. [8] motivates the need for a *device ontology* in the area of semantic service discovery. The proposed ontology gives a detailed description of a device and is separated into five categories that are based on the type of provided information: device description, hardware description, software description, device status, and service. This work does not specifically target mobile devices, but its principles are also applicable within the mobile domain.

The *PivOn Ontologies* [65] as well as the *Device Ontology* presented by the *Foundation for Intelligent Physical Agents (FIPA)* [53] are two other ontology examples focusing on a detailed representation of a device including its capabilities.

Villalonga et al. [150], on the other hand, propose the *Mobile Ontology* as a candidate for standardization that defines a semantic model for service delivery platforms in next generation networks. Several sub-ontologies, such as the *Services*,

Context, and *Distributed Communication Sphere (DCS)* vocabulary, are part of the *Mobile Ontology*. The *DCS Ontology*[26] provides a comprehensive collection of concepts describing a mobile device and its connectivity. It differentiates between GSM, UMTS, WiFi as well as Bluetooth networks and can further be utilized to specify network capabilities like QoS, network coverage, or delay rate.

Qiao et al. [120] present the *Telecommunications Service Domain Ontology (TSDO)* also consisting of several sub-ontologies, such as the *Terminal Capability Ontology* or *Network Ontology* [156], that semantically describe telecommunication network services. The *Network Ontology*, for instance, introduces concepts and features of mobile as well as fixed networks and models the relationship between different network access technology types such as GSM, UMTS or WiFi.

Based on their claim that there is a lack of existing consensual context models, Villalon et al. [119] introduce the *mIO! Ontology Network* that incorporates a set of vocabulary (e.g., *Device Ontology*, *Environment Ontology*, or *Network Ontology*) to be used in mobile environments and apply it in a paddle use case. Cleary et al. [40], on the other hand, address the complexity of configuring wireless networks and simplify this task by applying an ontology-based modeling approach in combination with workflows and reasoning.

Existing ontologies and models typically concentrate on describing technological aspects such as network connectivity, mobile devices, or a combination of both. They do not take topological information of mobile networks into consideration. Hence, the novelty of the *OpenMobileNetwork Ontology* stems from the semantic representation of mobile radio access networks and WiFi access points from a topological perspective by providing information about their positions, coverage areas, and neighboring cell relations. Another added value is that our ontology incorporates concepts for dynamic network context parameters such as the traffic and number of users within a mobile network cell.

2.5.4 User Profiles and Preferences

One essential part of user context comprises profile and preference information. A user profile consists of (mostly static) personal data about the user, such as his name, age, or social relationships, while preferences represent his likings, hobbies, and habits, for instance. Ontologies are also exploited for modeling this kind of information.

Probably, the most well-known vocabulary for representing user profiles and social relationships among users is *Friend of a Friend (FOAF)*[27] authored by Brickley and Miller [28]. Since its creation in 2000, it gradually evolved over time and provides two main categories of concepts. The terms of the core category "describe characteristics of people and social groups that are independent of time and technology" such as

[26]http://emb1.esilab.org/sofia/ontology/1.0/foundational/dcs.owl.

[27]*foaf*, http://xmlns.com/foaf/0.1/.

`foaf:Person` (standing for a person), `foaf:name` (e.g., representing the name of a person), or `foaf:knows` (describing a person known by another person). Information that is rather changing over time or due to technology, is included in the social Web category. Examples are online accounts (`foaf:OnlineAccount`) or things of interest to the user (`foaf:topic_interest`).

FOAF is extended by Davis [44] with the specification of the *RELATIONSHIP Vocabulary*.[28] This ontology provides more fine-grained concepts for relations such as family memberships (e.g., `rel:parentOf` or `rel:childOf`). Another extension to FOAF is made by Polo et al. [117]. The *RECommendations Ontology*[29] and later renamed *Framework for Ratings and Preferences Ontology*[30] enables the representation of user preferences with constraints and provides possibilities for clustering preference groups. The authors claim that their formal language makes it possible to exchange and use the preference data across various applications.

Apart from FOAF and its extensions, Golemati et al. [57] also propose a *user profile ontology* that focuses on modeling profile and preference information. The authors define concepts for a person, his characteristics, and abilities. Social relationships are represented via a *contact* concept, whereas preferences are separated into *preference*, *interest*, and *activity* terms. Heckmann et al. [63], on the other hand, present with *GUMO* a much more sophisticated *user model ontology* incorporating concepts for basic user dimensions such as the emotional and psychological state, characteristics, or personality of a user.

During the design process of the *Context Data Cloud Ontology*, several concepts of FOAF, the *RELATIONSHIP Vocabulary* as well as the *RECommendations Ontology* have been reused for describing the location and user context.

[28] *rel*, http://purl.org/vocab/relationship/.

[29] *reco*, http://purl.org/reco#.

[30] *frap*, http://purl.org/frap#.

Part II
Contribution

Chapter 3
Requirements

In this chapter, we build the basis for the thesis by highlighting the requirements for a semantically enriched mobile and WiFi network data platform that incorporates context information from external data sources. For this purpose, Sect. 3.1 gives a short overview about the application areas in which mobile network operators are able to offer semantically enriched context-aware services. Due to the fact that context information is very crucial for realizing such services, Sect. 3.2 discusses the requirements on context data based on the application areas and presents an analysis of the *LOD Cloud* in order to identify relevant context information in it. Finally, Sect. 3.3 lists the functional requirements of the platform, whereas Sect. 3.4 focuses on the non-functional requirements.

3.1 Application Areas

Semantically enriched mobile and WiFi network data in combination with other context data sources yields great potential to generate added value in the following application areas:

1. In-house services: Telecommunication providers can optimize their networks based on certain circumstances in order to reduce the overall costs, optimize the processes, the QoS, and the user-perceived QoE. One example is the development of a power management in mobile networks as introduced in Sect. 7.1, in which the current state of the network is analyzed in order to de- and reactivate mobile network cells based on network usage profiles. This leads to energy as well as cost savings. Section 7.1.1 describes a scenario and possible network optimization use cases that build the basis for identifying the context data requirements in Sect. 3.2.
2. B2C services: Besides all kinds of location-based services, mobile network operators can provide positioning methods comparable to or more sophisticated than the

A. Uzun, *Semantic Modeling and Enrichment of Mobile and WiFi Network Data*, T-Labs Series in Telecommunication Services,
https://doi.org/10.1007/978-3-319-90769-7_3

Google Maps Geolocation API[1] relying on very accurate mobile network topology data. The *Semantic Positioning* solutions showcased in Sect. 7.2, for example, improve classic geocoding as well as geofencing methods and add semantic features to proactive self-referencing and cross-referencing LBSs.

3. B2B services: Through well-defined interfaces, mobile network operators can provide analytics information to third-parties in a B2B relationship. Stores interested in the age groups of people passing by, for example, can get this information in order to adjust their product portfolio and marketing campaigns accordingly. See Sect. 7.3 for further details.

3.2 Context Data Requirements

Taking the application areas introduced in Sect. 3.1 as a reference with a special focus on the power management scenario and the network optimization use cases in Sect. 7.1.1, we identified two major sources including several types of context that support the realization of context-aware services within these areas [C3, C7]. Table 3.1 classifies context data incorporating context sources and their types, the corresponding context data parameters and a few examples.

The first context source (*network context*) comprises data originating from wireless radio networks and can be sub-categorized into different context types. The *static network context* or *a priori information context* type consists of static information parameters that are supplied by network providers or network component manufacturers and that are expected not to change often or anytime at all [81]. This type of context includes data describing the actual topology of the mobile network, such as the coordinates and coverage information of the base stations, or locations and ranges of WiFi access points. The *dynamic network context*, on the other hand, depicts constantly changing context information gained from network components, e.g., data describing the (average) traffic generated within a mobile network cell or the (expected) number of users at a given time. Another aspect of the dynamic network context incorporates *network user context* data acquired mainly from mobile devices. This type includes the location of the user, the list of mobile network cells in reach of him, the technical capabilities of his device (e.g., which radio technologies are supported), or information about service usage, for example.

Another context source is the *third-party context* that is usually acquired from external context providers (e.g., *LOD Cloud*, Web services, or databases) [29] in order enrich the information pool to be utilized for an enhanced context-aware service experience. In addition to the network optimization use cases in Sect. 7.1.1, the list of third-party context information is completed by also analyzing the data requirements of traditional context-aware services that are mainly part of the B2C or B2B service areas.

[1] https://developers.google.com/maps/documentation/geolocation/.

Table 3.1 Context Data Classification

Context Source	Context Type	Context Data	Examples
Network Context	Static Network Context (A Priori Information)	Base station locations	
		WiFi hotspot locations	
		Cell structure	Access technology, topology
		Antenna configuration	Tilt, orientation
		Hardware configuration	
		Manufacturer information	Power consumption
	Dynamic Network Context (incl. Network User Context)	Network traffic profiles	Average daily traffic
		Live network traffic	Current traffic, total traffic
		User activity	Number of active/idle users
		User position	Location, heading, speed
		User profiles	Age groups
		Cells in user proximity	Current cell, neighboring cells
		Service usage	HTTP, video streaming, apps
		Mobile device capabilities	Radio technologies, sensors
Third-party Context		Events	Time, place, duration, users
		Points of interest	Sights, restaurants, stadium
		Domain-specific information	Dishes of a restaurant
		Map data	Country, city, streets
		Car traffic	Congestions, rush hours
		Environment	Temperature, humidity, weather
		Calendar	Regional holidays
		User location preferences	Check-ins, ratings, comments
		Encyclopedia	Music, movies, sports

LBSs (e.g., location-based analytics, recommendations, reminders, advertisement, navigation, tracking, or gaming) are the most prominent class of CASs and have the highest demand on *location context* and its variants. Examples are POIs in combination with specific user location preferences (e.g., visiting frequency or popularity), events, or information about public transport systems. The location context is usually complemented by information describing the *environment* such as temperature, humidity, or more generically the weather condition [7].

Additional context data requirements arise from the area of context-aware recommendations [3, 78], where similarities or suitability for certain situations is computed by incorporating *encyclopedic* datasets with information about music, movies, or sports, for instance. Ostuni et al. [110] developed a context-aware movie recommendation system based on movie data from *DBpedia* [26] that also identifies theaters located nearby. These types of encyclopedic datasets can also be utilized as a valuable source for *disambiguation* or identification of *relatedness*. In the recommendation scenario, the similarity between the sport of handball and soccer, for example, can be identified by inferring that both are team sports in which goals are scored. This information can be used for recommending other team sport events.

Sophisticated services, such as an adaptive, context-aware and technology-comprehensive power management in mobile networks, can only be established if a combination of context data coming from the different context sources is exploited in a mashup-manner. As mentioned before, static network context is usually owned by mobile network operators and builds the basis for a power management in mobile networks since the network topology to be optimized has to be known. Dynamic network context, on the other hand, can either be used to show live information about the network (e.g., current traffic produced in a base station or number of current users in a cell) or to calculate several profiles using historic data such as network traffic profiles, service usage, or user movements. This type of context data can also be collected via smartphones using a device-centric crowdsourcing approach. Third-party context can be utilized to perform a more sophisticated power management that does not only take network parameters into consideration, but also other sources of information such as frequently visited points of interest or events taking place at a certain time in a stadium.

3.2.1 Analysis of Available Context Data in the LOD Cloud

Comparing the list of identified third-party context information in Table 3.1 and the data present in the *LOD Cloud*, we can see that a lot of the context information is already available in the cloud in the form of structured data.

The number of datasets within the *LOD Cloud* has been increasing rapidly [25] and has reached a total of 1,139 by February 2017 [2]. Since context-aware services within the application areas described in Sect. 3.1 mostly depend on a correlation of heterogeneous data sources, the huge pool of interlinkable context information available in the *LOD Cloud* can be exploited for enhancing the quality of the services.

Being able to retrieve context information in a uniform manner is another advantage of utilizing the data within the *LOD Cloud*. In contrast to other external data sources (e.g., Web APIs) where developers are forced to implement against proprietary APIs with different data formats and parse the returned results to the desired data model, the *LOD Cloud* provides data in a uniform format (namely RDF) that can be queried via the standardized SPARQL interface. This process reduces the manual implementation effort for extracting third-party context information to a great extent.

Based on the list of potential third-party context data, we performed an analysis of the *LOD Cloud* in [C12, J2]. We analyzed the applicability of the *LOD Cloud* as a (reliable) context information source and specified *discoverability* as well as *availability* of the linked datasets as crucial requirements for being applicable as a context provider. In addition, we investigated the actual contents of the datasets by approximating the content as well as the schema of some exemplary datasets.

3.2.1.1 Discoverability and Availability of Datasets

We identified *discoverability* as well as *availability* as relevant requirements, because with the ongoing growth of the *LOD Cloud*, it has increasingly become a challenging task to find datasets that incorporate relevant context information. Many researchers dealt with the problem of discoverability [61, 82, 149] and the *Vocabulary of Interlinked Datasets (VoID)*[2] [4, 162] was proposed as one solution for easing the dataset discovery. However, only a fraction of datasets published a VoID description to their datasets [30]. In addition, these descriptions are often too coarse for understanding the underlying data, which is also due to the fact that the concepts of VoID are rather of structural nature giving information about the number of triples or the dataset owner, for example.

For evaluating the discoverability of the datasets in the *LOD Cloud*, we worked on strategies to ideally compile a complete list of endpoints within the cloud. The *OpenLink LOD SPARQL endpoint*[3] as well as the *voiD Store*[4] are two repositories that aim to provide a collection of dataset endpoints that were annotated with VoID. Querying these datasets by making use of the `void:sparqlEndpoint` concept, returns only a small number of endpoint addresses showing that these repositories do not encapsulate all linked datasets and that the majority of the datasets do not incorporate a VoID description. This leads to the conclusion that VoID is currently not a reliable source for discovering a complete list of datasets in the *LOD Cloud*.

Another alternative is *datahub*[5] offering meta data for a broad collection of datasets published in *Linked Data* format, where we retrieved 439 endpoint addresses by the time of our investigations in February 2014. By implementing several scripts, we analyzed the availability of the 439 endpoints and found out that typically 53%

[2] *void*, http://rdfs.org/ns/void#.

[3] http://lod.openlinksw.com/sparql.

[4] http://void.rkbexplorer.com/sparql/.

[5] https://datahub.io/organization/lodcloud.

to 58% of the 439 tested endpoints responded with `200 - OK` for multiple test runs distributed over a month time period. The other endpoints were not reachable one way or the other, i.e., the endpoint address could not be resolved, a timeout occurred, or a socket exception was raised. Another interesting aspect was that 7% of all endpoints returned a `401 - Unauthorized` or `403 - Forbidden` status code meaning that the dataset owners do not want their data to be publicly available even though they are listed at *datahub*. These datasets might follow a similar approach as discussed in Sect. 4.2.1 by making use of data within the *LOD Cloud*, but closing their data to the public.

3.2.1.2 Contents of Datasets

After having evaluated the discoverability as well as the availability, we performed an analysis of the actual contents of the datasets that were accessible. Here, we applied two different methodologies.

At first, we tried to approximate the content of a specific dataset by having a deeper look into its concepts. We used the query in Listing 3.1 in order to retrieve the *30 most often* used *properties* and *classes* with the assumption that the concepts with the highest counts are the predominant types of the dataset incorporating information about the underlying data.

```
SELECT DISTINCT ?type (COUNT(?type) AS ?count)
{
  ?s a ?type .
}
ORDER BY DESC(?count) LIMIT 30
```

Listing 3.1 Query to retrieve the 30 most often used types

Executing the query for the datasets of *DBpedia* and *LinkedGeoData*, for example, returned the results listed in Table 3.2. We can see that the more generic and extensive a dataset is (e.g., *DBpedia*), the less specific are the results. A look at the 30 most often used concepts of *DBpedia* does not provide a clear understanding of the dataset contents. However, a highly specific dataset, such as *LinkedGeoData*, gives us a clue about the underlying data. The results of LGD clearly indicate that this dataset is geo-related with `geom:Geometry`, `spatial:Feature`, or `lgdo:Amenity`[6]being one of the highest counted concepts.

However, a drawback of this approach is that some non-informative concepts with a rather structural nature (e.g., `owl:Thing` or `rdf:type`), which do not provide any added value for understanding the contents, manipulate the results due to their extensive usage within the datasets. This problem can be overcome by (post-) filtering these types. In addition, this methodology could result in a list with several synonymous types that originate from different vocabularies (e.g., `foaf:Person`

[6] LinkedGeoData Ontology, *lgdo*, http://linkedgeodata.org/ontology/.

Table 3.2 List of 30 most often used types for *DBpedia* and *LinkedGeoData*, Dec 29th, 2016

Count	DBpedia	LinkedGeoData
1	foaf:Document	geom:Geometry
2	owl:Thing	spatial:Feature
3	yago:PhysicalEntity100001930	lgdm:Node
4	dbo:Image	lgdm:Way
5	yago:Object100002684	lgdo:PowerThing
6	yago:YagoLegalActorGeo	lgdo:Amenity
7	yago:Whole100003553	lgdo:PowerTower
8	dul:Agent	lgdo:BarrierThing
9	dbo:Agent	lgdo:Place
10	yago:YagoPermanentlyLocatedEntity	lgdo:RailwayThing
11	yago:YagoLegalActor	lgdo:PowerPole
12	foaf:Person	lgdo:Building
13	dbo:Person	lgdo:Parking
14	wikidata:Q5	lgdo:Fence
15	schema:Person	lgdo:ManMadeThing
16	wikidata:Q215627	lgdo:Shop
17	dul:NaturalPerson	lgdo:Rail
18	skos:Concept	lgdo:UnclassifiedBuilding
19	yago:LivingThing100004258	lgdo:HighwayThing
20	yago:Organism100004475	lgdo:Wall
21	yago:CausalAgent100007347	lgdo:TourismThing
22	yago:Person100007846	lgdo:Locality
23	yago:Abstraction100002137	lgdo:Boundary
24	dbo:TimePeriod	lgdo:Village
25	yago:YagoGeoEntity	lgdo:Hamlet
26	geo:SpatialThing	lgdo:Leisure
27	dbo:CareerStation	lgdo:SportThing
28	dbo:Location	lgdo:PublicTransportThing
29	schema:Place	lgdo:Gate
30	dbo:Place	lgdo:AdministrativeBoundary

and `dbo:Person`). Here, ontology matching and alignment techniques [75, 102] could be applied in order to map both types to the same concept. Finally, executing a query as given in Listing 3.1 turns out to be computationally costly, which leads to timeouts at a lot of working endpoints.

The content approximation methodology is rather simple and provides information on an abstract level (also due to the limitations mentioned above). In order to get more insight about the content of a dataset, it is of importance to understand the schema or ontology including the class hierarchy and the properties of each class.

Ideally, this would not be necessary if ontology specifications were provided alongside to the published datasets. However, this is often not the case even though these specifications make it possible to comprehend the contents of the datasets.

By using a set of SPARQL queries with the `CONSTRUCT` function that we have assembled, we were able to approximate the underlying schema of some endpoints including all classes, the class hierarchy, and all object as well as datatype properties. However, many endpoint maintainers limit the execution time of a query, so that attacks are not possible. Therefore, this approach could not be processed extensively.

Utilizing both content approximation methods as much as possible, we categorized datasets that were of particular interest for enriching CASs in general and indirectly in the telecommunications domain. Due to the sheer size of the *LOD Cloud* and the variety of data, it is not possible to cover all application areas nor domains. Therefore, a coarse categorization is given with a few context data examples for each bullet point:

- **Location**: POIs, geographic information
- **Environment:** weather, traffic, sensors (e.g., environmental, meteorological, and physical)
- **Media**: music, albums, genres, artists, movies, actors, directors, studios
- **Meta**: legal information
- **Open government data**: organizations, institutions, companies
- **Education**: courses, accreditation, institutes
- **Bibliographic data**: research areas, books, magazines, publications, theses, authors, newspapers, audio, images
- **Biological, chemical, medical data**: genes, diseases, species, drugs, enzymes, chemical, structures, biological pathways
- **Language**: natural language, lexicographical information, translations
- **Public data**: regulations, amendments, decisions, health data, wards, ministries
- **Regional archives**: diverse archived data, places, books, language

Table 3.3 lists a set of exemplary dataset endpoints (available by the time of our studies) being viable context providers that cover some of the aforementioned domains in particular depth or contain additional information that can be utilized.

The deep dive into the *LOD Cloud* shows that some of the third-party context information, which we have identified to be of particular interest for our three application areas (e.g., POIs, map data, sensor data, or encyclopedic information) are already available in the cloud in a structured form. Even though some other data is still missing (e.g., user location preferences or events), the covered domains and the interlinking nature of the data emphasize that the *LOD Cloud* can be a valuable third-party context source for semantically enriching and thus improving the QoE of context-aware services based on mobile and WiFi network data.

Table 3.3 Exemplary LOD Datasets and Covered Domains, Feb 2014

Endpoint address	Description
http://linkedgeodata.org/sparql/	Geo-related data including POIs
http://sonicbanana.cs.wright.edu:8890/sparql	Sensor data including weather
http://environment.data.gov.uk/sparql/bwq/queryv	Bathing water quality
http://aemet.linkeddata.es/sparql	Meteorological data
https://dbpedia.org/sparql	Encyclopedic data
http://lod.b3kat.de/sparql	Documents, scientific publications, software
http://bnb.data.bl.uk/sparql	British library
http://dbtune.org/jamendo/sparql/	Music data
http://api.talis.com/stores/musicbrainz/services/sparql	Music data
http://data.linkedmdb.org/sparql	Movie data
http://data.linkedu.eu/kis/query	Educational courses and institutes
http://linked.opendata.cz/sparql	Czech open data
http://prelex.publicdata.eu/sparql	European Union inter-institutional law

3.3 Functional Requirements

This section lists the functional requirements for a platform offering semantically enriched mobile network and WiFi access point topology data based on the principles of *Linked Data* with interlinks to other datasets of the *LOD Cloud*. The defined requirements in the following subsections are based on the final design decisions that are made in Chaps. 4 and 5, i.e., we assume that we need to collect mobile and WiFi network data as well as additional third-party context information via crowdsourcing in order to have datasets to work with.

3.3.1 Network Context Data Platform

1. The platform grants access to relevant network context data (including static as well as dynamic network context) as defined in Sect. 3.2 via a SPARQL endpoint.
   a. The platform provides access to geographically mapped mobile network (topology) data from potentially all telecommunication providers worldwide comprising cell positions and coverage areas, neighboring cell relations, access technologies, cell network traffic, user activity as well as service usage data, and mobile device capabilities.
   b. The platform provides access to geographically mapped WiFi network (topology) data from potentially all access points worldwide including their positions and coverage areas.

2. The platform provides an HTML as well as RDF representation of the network context data.
3. The platform provides a visualization of the network context data in the form of a coverage map.
    a. For Germany, the map provides an operator-specific visualization by separating the cells of each operator by different colored markers. Outside of Germany, all cells are represented by a standard colored marker.
    b. The map provides several display options for filtering the cells based on access technologies or operators, for illustrating the coverage areas as a circle or polygon, and for additionally showing the WiFi access points.
    c. WiFi access points are represented by a standard colored marker.
    d. Clicking on one of the cell (or WiFi AP) markers illustrates the coverage area of the cell (or WiFi AP) and opens a tab with additional information such as the related RDF data, the neighbor relations, network traffic as well as service usage information, and a list of devices being connected to the cell (or WiFi AP).
4. The platform provides proof of concept implementations of semantically enriched context-aware services for each application area as identified in Sect. 3.1.

3.3.2 Network Context Data Processing

1. The measurement client acquires network context data (including static as well as dynamic network context) as defined in Sect. 3.2 via crowdsourcing.
    a. The measurement client collects accurate information about the cell the mobile device is connected to. This information comprises the Cell-ID, the LAC, the MCC, the MNC, the network access type, the *Received Signal Strength Indicator (RSSI)*, the *Primary Scrambling Code (PSC)*, and the operator name.
    b. The measurement client acquires accurate information about nearby cells including the Cell-ID, LAC, PSC, the network access type, and RSSI.
    c. The measurement client gathers accurate WiFi network data incorporating the *Basic Service Set Identification (BSSID)*, the *Service Set Identifier (SSID)*, and RSSI.
    d. The measurement client collects accurate information about the mobile device consisting of its location during the measurement given in WGS84 coordinates, the accuracy of the location information, a timestamp when the data was collected, the operating system, the device model and hardware name, the manufacturer name, information about the SIM card, the roaming state, the traffic consumption of the device, and the mobile services used.
    e. The measurement client conducts the network measurements in the background, so that the user does not need to keep the app in the foreground.

2. The measurement client sends the collected data in a certain time interval to the server.
    a. The measurement client checks if the device is online. If it is not online, it stores the network measurements on the internal device storage. If the device is online, the measurements are transferred directly to the server.
    b. Measurements that have been stored locally are uploaded manually at anytime when the device is online again.
3. Based on the set of crowdsourced network measurements, the platform very accurately approximates the network topologies (including the positions and coverage areas) of potentially all telecommunication providers and WiFi access points worldwide.

3.3.3 Third-Party Context Dataset

1. Via a SPARQL endpoint, the third-party context dataset provides access to third-party context information as defined in Sect. 3.2 that is relevant for the semantically enriched services within the application areas, but is not yet available in the *LOD Cloud*.
2. The third-party context dataset provides an HTML as well as RDF representation of the third-party context information.

3.3.4 Third-Party Context Data Processing

1. The third-party context collection client acquires third-party context as defined in Sect. 3.2, which is not yet available in the *LOD Cloud*.
    a. The client gathers dynamic information about the popularity of certain places or the visiting frequency of users in specific contextual situations, e.g., check-in and check-out times, the weather condition by the time of a check-in, whether the check-in occurred during a regional holiday, whether the visit happened in an indoor or outdoor environment, location preferences in the form of favorite or frequently visited locations, and domain-specific information such as dishes of a restaurant.
    b. The client collects the data whenever a user performs a check-in or a check-out.
2. The third-party context collection client sends the collected data to the server whenever a user performs a check-in.
3. The third-party context collection client sends an update of the collected data whenever a user performs a check-out.

3.4 Non-functional Requirements

Complementing the functional requirements in Sect. 3.3, this section gives an overview about the non-functional requirements of the platform. These requirements mainly focus on modeling, publishing, and interlinking the network and third-party context data according to the principles of *Linked Data*.

1. Network context data (including static as well as dynamic network context) as defined in Sect. 3.2 is modeled and published in *Linked Data* format.
   a. Mobile network (topology) data is modeled and published according to the principles of *Linked Data*.
   b. WiFi network (topology) data is modeled and published based on the principles of *Linked Data*.
   c. The ontology describing network topology data reuses well-known vocabulary concepts.
   d. A meta description of the network topology dataset is given using VoID.
2. The network topology data is interlinked to third-party context information as identified in Sect. 3.2 that is already available in the *LOD Cloud* (e.g., points of interest or encyclopedic data).
3. Third-party context that is relevant for the semantically enriched services within the application areas, but is not yet available (e.g., user location preferences or domain-specific information) in the *LOD Cloud*, is modeled and published according to the principles of *Linked Data*.
   a. The third-party context data is modeled and published in *Linked Data* format.
   b. The ontology specifying the third-party context reuses well-known vocabulary concepts.
   c. The created context dataset is interlinked to the network topology data.

Chapter 4
Semantic Enrichment of Mobile and WiFi Network Data

Chapter 4 comprises the main contribution of this thesis and sheds light on the process of semantically enriching mobile and WiFi network data. As a result of this process, it introduces the *OpenMobileNetwork* [C7, J1], which is an open platform for providing approximated and semantically enriched mobile network and WiFi access point topology data based on the principles of *Linked Data* [62]. Network measurements that are constantly collected via crowdsourcing are used to infer the topology of mobile networks and WiFi access points worldwide. The created dataset is further interlinked with other datasets in the *LOD Cloud* (e.g., *LinkedGeoData* or *Linked Crowdsourced Data*) for providing additional context information that is geographically and semantically related to the network topology data such as POIs covered by the reception areas of the mobile network cells. It is made available to the public through a SPARQL endpoint[1] and described using VoID.[2] With thc update of the diagram in August 2014 (see Fig. 4.1), the *OpenMobileNetwork* officially became part of the *LOD Cloud*.

Section 4.1 highlights the data sources providing mobile network and WiFi AP data, whereas Sect. 4.2 defines a functional architecture for the *OpenMobileNetwork* by analyzing alternative architectures. In Sect. 4.3, on the other hand, several crowdsourcing methods are discussed that are utilized in order to collect network context data for the platform, while Sect. 4.4 illustrates how network topologies are estimated out of a set of collected measurements. Finally, Sect. 4.5 presents the "semantification" process of the mobile and WiFi network data by describing the *OpenMobileNetwork Ontology* in detail.

[1] http://sparql.openmobilenetwork.org/.

[2] http://www.openmobilenetwork.org/resource/void/.

A. Uzun, *Semantic Modeling and Enrichment of Mobile and WiFi Network Data*, T-Labs Series in Telecommunication Services,
https://doi.org/10.1007/978-3-319-90769-7_4

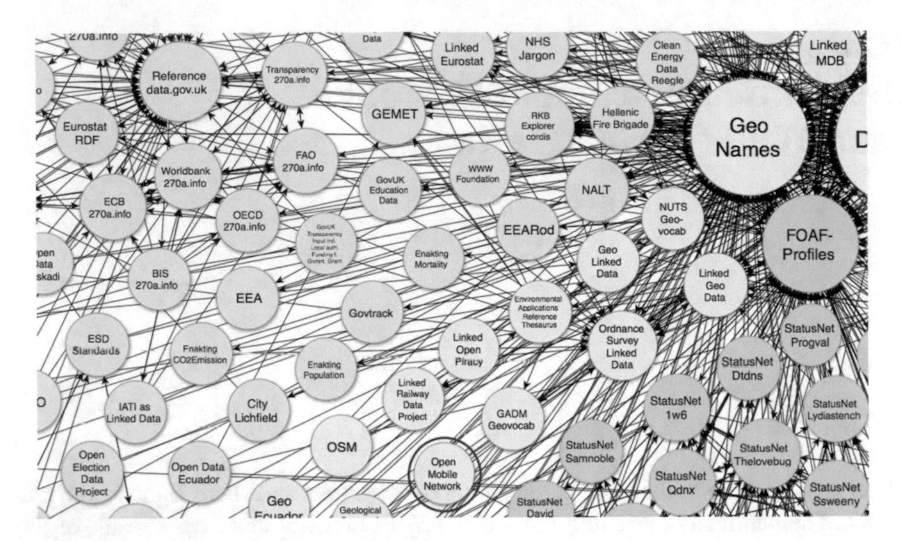

Fig. 4.1 OpenMobileNetwork in the LOD Cloud

4.1 Network Data Sources

In this section, we present several data sources comprising information that enable the provision of context-aware services based on wireless network data. Section 4.1.1 gives an overview about valuable data that mobile network operators possess, whereas Sect. 4.1.2 presents cell and WiFi databases that are (partly) publicly available on the Web. We conclude each section with a discussion about the decision of whether these data sources are suitable for our purposes or not.

4.1.1 Mobile Network Data

There is a huge variety of data that is available within mobile networks and that telecommunication providers extract out of different components of the mobile network architecture. In [107], the authors present a detailed overview about what kind of information is available in mobile networks and out of which component to collect it.

Data collected in the radio access network of a mobile network (e.g., GSM, UMTS, or LTE) comprises information that is related to mobile devices such as "network attach and detach operations, start and conclusion of sessions, handover, location update events related to any call, texting, or data transfer activity" or the "uplink and downlink throughput experienced by the device".

In the core network, on the other hand, there is plenty of subscriber data available in different components [48]. The available information depends on whether the data is extracted out of components in the circuit-switched or packet-switched core domain (including the EPC). The MSC as part of the CS core domain, for example, provides data that is similar to the information collected in the RAN, but only for events related to voice and texting with a granularity on location area level (comprising several BSCs with several connected base stations to each of them). IP traffic that is transmitted and received by mobile devices, is extracted out of the GGSN or PDN-GW components including "IP session start and end time, device and user identifiers, traffic volume, type of service", and service classes with "associated location information". In addition, *Packet Data Protocol (PDP)* context messages that are sent by the SGSN or MME to the GGSN or PDN-GW in order to establish, update, and close IP sessions of the mobile devices, include the cell the user is connected to when using an IP session.

Another component is the *Charging Gateway Function (CGF)* that provides *Call Detail Records (CDR)*. Based on these CDRs, the network operator determines the fees to be charged to his users. The CDRs usually contain the "start timestamp, duration, and originating cell sector of each voice, texting and data traffic activity of every device". Concrete examples of CDR parameters are *Short Message Service (SMS)* and calls initiated from a mobile device, SMS and calls received on a device, location update after timer expired or LAC of the mobile device has changed, and information about a handover after a mobile device initiated or received a call.

Besides the data that is generated within mobile networks, information about the topology and infrastructure of the networks is usually available offline in the form of manually maintained lists. Among other things, these lists contain the Cell-IDs, LACs, geo coordinates of the base stations, and their coverage areas given as polygons, for example.

As we can see, telecommunication providers possess all kinds of mobile network data and also have the possibility to extract all of it in the desired granularity for providing new and innovative context-aware services (as companies like *MotionLogic* or *Telefónica* [15] already do). The variety and mass of information enables network operators to implement various application scenarios (as introduced in Sect. 3.1 and Chap. 7). Nevertheless, the full potential of network data exploitation is still restricted due to the fact that the integration of semantics into the services by enriching mobile network data with semantic technologies as well as context data (available within the *LOD Cloud*), is not considered yet.

In order to showcase a proof of concept implementation of a platform with semantically enriched mobile network data that is interlinked with other context information sources, we need to have a dataset of mobile network data to work with. However, we were not able to get real data from an operator due to many reasons: First, this kind of data is kept very secret as it is treated as the main asset of the company, which contradicts to the idea of the *LOD Cloud* with data being open to anybody and shared all over the Web. Another aspect is that operators are obliged to protect the privacy of their customers, which often hinders them to share such sensitive data even for research purposes.

Table 4.1 Overview of cell and WiFi AP databases, Oct 4th, 2015

Commercial Cell and WiFi AP Databases

Database	Cells	Operators	WiFi APs	Countries
Google	NA	Worldwide	NA	Worldwide
Apple[a]	NA	Worldwide	NA	Worldwide
Skyhook Wireless[b]	NA	NA	NA	NA
Combain[c]	>65M	>1000	>749M	196
Navizon[d]	NA	NA	NA	NA
OpenSignal[e]	824,297	825	>1.2B	Worldwide
Cellspotting[f]	120,453	667	NA	NA

Open Cell and WiFi AP Databases

Database	Cells	Operators	WiFi APs	Countries
OpenBMap	492,486	NA	4.3M	NA
OpenCellID	>7M	NA	NA	NA
RadioRaiders Cellumap[g]	NA	NA	NA	NA
Mozilla Location Service[h]	15.48M	NA	380.61M	223

[a]https://developer.apple.com/library/ios/documentation/UserExperience/Conceptual/LocationAwarenessPG/
[b]http://www.skyhookwireless.com/
[c]http://www.location-api.com/
[d]http://www.navizon.com/navizon-how-it-works
[e]https://www.opensignal.com/
[f]http://www.cellspotting.com/
[g]http://www.cellumap.com/
[h]https://location.services.mozilla.com/

4.1.2 Cell and WiFi AP Databases

Due to the fact that we were not able to receive real mobile network data from an operator, we analyzed several cell and WiFi AP databases that are available on the Web and evaluated whether they are suitable for the implementation of our proof of concept platform and the services. Table 4.1 gives an overview about commercial as well as open cell and WiFi AP databases. A similar list of cell databases is maintained at *Wikipedia*.[3]

Realizing the strategic potential of mobile network and WiFi AP data very early, *Google* was the first company that implemented a cell and WiFi AP database as a commercial solution based on data collected over the years from millions of

[3]https://en.wikipedia.org/wiki/Cell_ID.

crowdsourcing devices and *Google Street View* cars. At first, this information has been used as a competing positioning solution to GPS within their *Google Maps Mobile* application [39]. Currently, it is utilized in a manifold way within their different Location APIs such as the *Google Maps Geolocation API*, the *Google Places API*, or the *Network Location Provider* as part of the *Android LocationManager*.[4]

Other platform providers, such as *Apple* [141] or *Microsoft* [103], first used external location providers (e.g., *Google*, *Skyhook Wireless*, or *Navizon*) for their services, but also moved on to implementing their own cell and WiFi AP databases due to its strategic relevancy.

Due to the fact that this data intrinsically enables the operators to receive the indirect whereabouts of their users as a valuable asset for different kinds of LBSs, none of the commercial database owners as seen in Table 4.1 are willing to share the data in its completeness. Only fragments of this data are provided through APIs, which are necessary for fulfilling the functionality of the offered services such as the returned WGS84 coordinates for a positioning request containing cellular identifiers, WiFi AP information, and optional RSSI.

As a countermovement to the commercial providers, completely open databases arose over time, which allowed access to the entire available data collected by their own, but much smaller crowdsourcing communities.

OpenCellID (see footnote 6 in Chap. 1), for example, is the most famous open data project in this area led by *Unwired Labs*. It provides cell data for countries all over the world under the *Creative Commons Attribution-ShareAlike 3.0 Unported (CC BY-SA 3.0)*[5] license. As of Jan 2015, the database consisted of more than 7 million cells calculated out of more than 1.2 billion measurements. The data is provided either via API or in the form of a CSV dump. Smartphone clients for various operating systems[6] can be downloaded in order to further extend the database with cell information.

OpenBMap (see footnote 7 in Chap. 1), on the other hand, is an alternative project under the *Creative Commons Attribution-Share Alike 3.0 Unported* and *Open Database License (ODbL) v1.0*[7] licenses and collects not only mobile network, but also WiFi AP and Bluetooth data. By October 4th, 2015, *OpenBMap* has been acquiring 492,486 cells and 4.3 million WiFi access points worldwide.

Unfortunately, by the time of our first studies in 2012 and 2013, both projects lacked network measurement parameters that were necessary for the realization of our application scenarios. *OpenCellID*, for example, did not provide any information about the type of a cell, so that it was not possible to distinguish between 2G (GSM, GPRS, EDGE) and 3G (UMTS, HSPA) cells. Furthermore, no information was given about neighboring cells of a serving base station. In addition, the cell range attribute was mostly `0`, which made this parameter almost useless.

[4]http://developer.android.com/reference/android/location/LocationManager.html.

[5]http://creativecommons.org/licenses/by-sa/3.0/.

[6]http://wiki.opencellid.org/wiki/Data_sources.

[7]http://www.openbmap.org/openBmap-wifi-odbl-10.txt.

Table 4.2 Comparison of network measurement parameters, June 2013

Attribute	Description	OpenCellID	OpenBMap	OpenMobileNetwork
mcc	Mobile Country Code	X	X	X
mnc	Mobile Network Code	X	X	X
lac	Location Area Code	X	X	X
cellid	Cell-ID	X	X	X
lat, lng	Latitude, Longitude	X	X	X
acc	Accuracy of GPS Coordinates			X
range	Cell Coverage	(X)		X
act	Network Access Type		X	X
neighbor	Neighboring Cell Information		X	X
ss	Signal Strength	X	X	X
rxlev	Reception Level (Optional)		X	
ta	Timing Advance		(X)	
measures	No. of Measurements per Cell	X		X
sim-operator	SIM Card Operator			X
psc	Primary Scrambling Code		X	X
roaming	Roaming State			X
traffic	RX/TX Total Bytes			X
model	Phone Model and OS		X	X
service	Service Usage Information			X

In contrast to *OpenCellID*, *OpenBMap* provided much more valuable information, such as the network access type and the neighboring cells (even though the neighboring cell information was only available for 2G networks), but paled in comparison to *OpenCellID* in terms of the number of collected measurements and therefore worldwide coverage.

Dynamic aspects of a mobile network (e.g., the current traffic and the number of users in a mobile network cell or service usage information) are not utilized by any of those open databases even though the historic and live state of a mobile network cell is of utter importance when realizing sophisticated and innovative context-aware services based on mobile network data. Table 4.2 lists the parameters fulfilling the application scenario requirements and compares the raw data pool of the open data projects to the dataset of the *OpenMobileNetwork*. Here, *X* represents parameters being conducted, whereas *(X)* are attributes that are listed on the respective project websites as being collected, but with no valid values available within the raw datasets.

Due to the mentioned reasons above as well as the fact that there is no quality evaluation of the cell data available, we decided to build our own open data platform by collecting network topology measurements in combination with dynamic as well as live network context data for establishing a complete dataset of mobile network cells and WiFi access point locations as well as associated coverage areas, neighboring cell relations, and live network data.

4.2 Architecture

This section gives insight into the process of designing a functional architecture for the *OpenMobileNetwork* by discussing the advantages and drawbacks of several architectural alternatives in Sect. 4.2.1. Section 4.2.2, on the other hand, presents the final functional architecture and describes its relevant components.

4.2.1 Architectural Alternatives

There are manifold ways of designing a functional architecture for an open data platform providing semantically enriched network context data. The design decisions are mostly influenced by the fact whether (real) mobile network data is made available by operators and whether in future telecommunication providers are willing to collaborate and share their data in order to compete against other service providers.

In this section, some architectural alternatives are introduced and the advantages as well as drawbacks are discussed. These exemplary architectures solely focus on the functional entities that are relevant for the design decisions and do not incorporate all platform functions.

Figure 4.2 illustrates the most straight-forward and simple approach with one central platform and a triplestore. Here, network context data (including mobile network cells as well as WiFi access points) is collected by crowdsourcing relevant network context information from smartphone users. The acquired data is processed as well as semantified by a *Network Context Data Processing* function of the central *OpenMobileNetwork* platform and stored into its triplestore. The semantification process is based on the *OpenMobileNetwork Ontology* (see Sect. 4.5.1) that represents the semantic description of the network topology dataset and that defines links to other datasets within the *LOD Cloud*.

This approach provides one endpoint to the *entire* semantic dataset ideally consisting of static as well as dynamic mobile network data of all providers and WiFi access points worldwide. Having an entire dataset in one triplestore facilitates the access and interaction with the data. Internal dataset links between mobile network

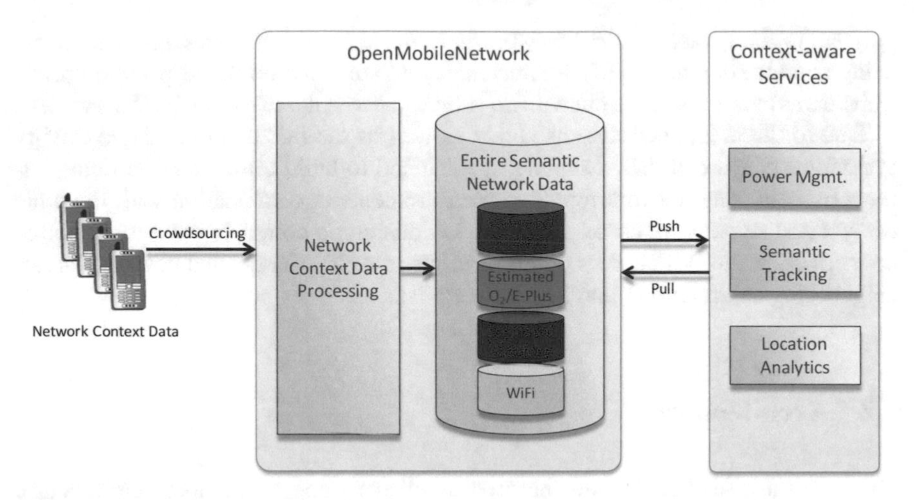

Fig. 4.2 Architectural alternative – central crowdsourcing platform

cells (based on their geographic topology, different access technologies, and operators) as well as between mobile cells and WiFi APs are easily defined and queried via SPARQL for the use within context-aware services. In addition, from an *LOD Cloud* perspective, links to and from other datasets are also created in a straight-forward fashion for the OMN as well as external LOD dataset owners.

However, the main drawback of this architecture is that it does not rely on real network topology information from the network providers, but rather on crowdsourced and estimated data collected via smartphones. The data quantity as well as quality is strongly dependent from the crowdsourcing user base meaning that the more users contribute with their data, the more the OMN will be able to provide a more or less complete network topology map worldwide. Furthermore, the collected data will always be an approximation of the real network topologies of the operators where different parameters, such as the number or the distribution of the measurements, will decide over the quality of the estimations. Taking common user movement patterns as a reference, network measurements will usually be available at high-frequented areas, whereas small streets or parks, for example, will not be covered at all. This inhomogeneous distribution will directly affect the approximation of the network topologies, which will then have an influence on the quality and accuracy of the provided context-aware services.

Another drawback is that the crowdsourcing part of the OMN is solely implemented in order to have a dataset of static and dynamic network context data to work with. Usually, operators already possess this data in a much more accurate version. So, if there was the possibility of accessing their data, we could save time and effort for realizing crowdsourcing mechanisms for the OMN.

As mentioned above, when providing context-aware services based on mobile network data, high quality and accuracy is mostly achieved if real network topology data is used as a basis. However, telecommunication providers usually view their

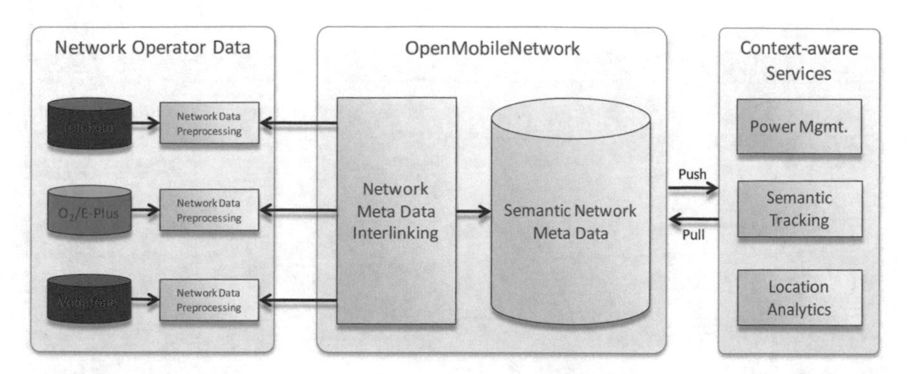

Fig. 4.3 Architectural alternative – network meta data interlinking

mobile network data as their asset and therefore keep it very secret. They are not willing to share it on a central platform or as part of the *LOD Cloud*. In order to ensure data security on the operator side, the OMN could act as a layer for interlinking the operator-specific data on a meta level with the data of other providers as well as with external datasets of the *LOD Cloud*, which is depicted in Fig. 4.3.

Within this approach, every network operator connected to the OMN runs an own database with preprocessed mobile network data that is compatible to the network topology description of the *OpenMobileNetwork Ontology* (see Sect. 4.5.1). This database incorporates information about the base station positions, the mobile network cells running on the base stations, their coverage areas defined as geographic polygons, and information about the neighboring cells, for example. The *Network Meta Data Interlinking* function processes the operator's database oncc and creates RDF links on a meta level between neighboring mobile network cells as well as between cells and external datasets (e.g., POIs) using the geographic topology information given within the database. These RDF statements are stored in a central OMN triplestore and do not provide more information than a link from one cell to another or from a cell to a POI, for example. Detailed information about the cell, such as its position or coverage area, is only given as a reference link to the operator's closed database on which only he has access to. External service developers or other telecommunication providers only work with the meta level links that provide enough semantic information to offer context-aware services.

The advantage of this approach is that it keeps the data of the network providers on their own databases, but nevertheless enables its exploitation for context-aware services. In order to build a *restaurant recommender* service, for example, it is completely sufficient to have the information that a certain mobile network cell covers a certain restaurant (e.g., `omn-owl:Cell omn-owl:coversLocationEntity cdc-owl:LocationEntity`, see Sect. 4.5.1). Whenever the user enters this cell, he will get notified of the restaurant in his vicinity. Detailed information about the exact position of the cell or its coverage area is not needed on the application development side.

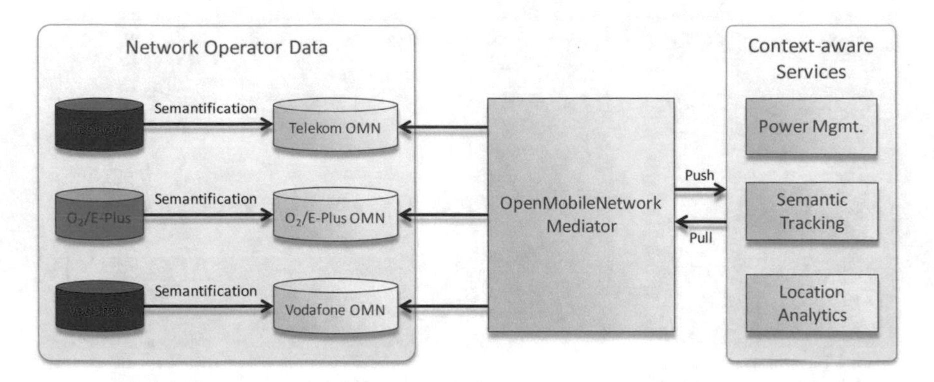

Fig. 4.4 Architectural alternative – data federation

However, some developers could try to "reengineer" mobile network topologies (including the positions and coverage areas of mobile network cells) by mapping the locations of the POIs (which are publicly available) to the cells they are covered from. But, this "reengineering" attempt will also only deliver estimations of the network topologies similar to the crowdsourcing approach discussed above, so that the real data of the network operators will still be protected.

Another drawback of this approach is that it is unlikely to get every operator worldwide connected to the OMN using such an approach. This, however, will lead to the fact that context-aware services will only work in regions where network providers are using the OMN. One way to tackle this problem could be to apply a hybrid approach in which operators could connect with their real data to the OMN and the remaining network topologies of the other operators could be estimated and added to the platform via crowdsourcing. This approach would also be applied for WiFi AP data.

The architectural alternatives in Figs. 4.2 and 4.3 are based on the assumption that the OMN is an independent centralized platform and mobile network data is shared (to some degree) on a central triplestore. However, network providers might want to keep everything in their closed environment and might not even want to share data on a meta level. Nevertheless, they might want to make use of the OMN features and semantically enriched context-aware services. Therefore, in our final example that is depicted in Fig. 4.4, we illustrate a data federation approach with multiple OMN instances for each network provider.

Within this approach, every operator runs his own OMN instance on his premises that incorporates a semantically enriched version of his own mobile network data including links to external datasets from the *LOD Cloud*. The operator can decide whether he wants to provide context-aware services that solely work for his customers. If this is the case, he uses the SPARQL endpoint of his own OMN instance in order to respond to service queries. However, if he wants to make sure that the provided context-aware services work operator-independent, the *OMN Mediator* acts as the endpoint that receives the query and forwards this query to the right OMN instance.

This architecture has the advantage that an operator-specific OMN instance enables a network provider to utilize the full potential of the OMN by keeping his network data completely closed and on his premises. It gives him the opportunity to decide whether he wants to provide network-specific or network-independent context-aware services using the *OMN Mediator*. The service queries that the *OMN Mediator* forwards to the OMN instances are completely hidden to the external service developer or other telecommunication providers.

However, this approach does not fully satisfy the idea of the *LOD Cloud*. Even though, the OMN instances store and link data based on the principles of *Linked Data*, the interlinking is only done one-way from the network provider to external datasets. Links from the *LOD Cloud* to the OMN cannot be established since the OMN is not (publicly) available in the *LOD Cloud*. This restricts the usage possibilities of the dataset for other *Linked Data* poviders.

Another aspect is that inter-operator links between mobile network cells are not possible with this architectural approach, i.e., cells of various operators that overlap each other geographically, for example, cannot be linked to each other, because every operator runs an own OMN silo for himself. In addition, as discussed above, this architecture design should also be combined with crowdsourcing functions as it is unlikely that all operators worldwide will be connected to the OMN.

4.2.2 Functional Architecture

For our functional architecture, we decided to apply the crowdsourcing solution due to several reasons: First, we were not able to get real network topology data of an operator to use within our platform, so that we were more or less "forced" to apply crowdsourcing in order to have a dataset to work with. Secondly, this approach enabled us to be flexible in selecting what kind of data to collect and in designing the data model for our purposes. Furthermore, we wanted to be part of the LOD community contributing with our dataset to the *LOD Cloud* in order to create awareness and show the potential of context-aware services based on mobile and WiFi network data.

Figure 4.5 illustrates the functional architecture of the *OpenMobileNetwork* that consists of several functional entities.

Network context data is collected via several crowdsourcing apps running on mobile devices that deliver the data in the form of network measurements to the OMN. These raw measurements are sent to the *Measurement Data Manager (MDM)* function, which provides several communication interfaces (*Client Connectors*) for the respective apps, e.g., *OMNApp* or *Jewel Chaser*. The incoming data is filtered according to certain constraints in order to ensure the accuracy of the measurements. For example, the *Client Connector* checks if the measurement values are within a feasible range by matching the geographic location and the MCC [73] of the cell to

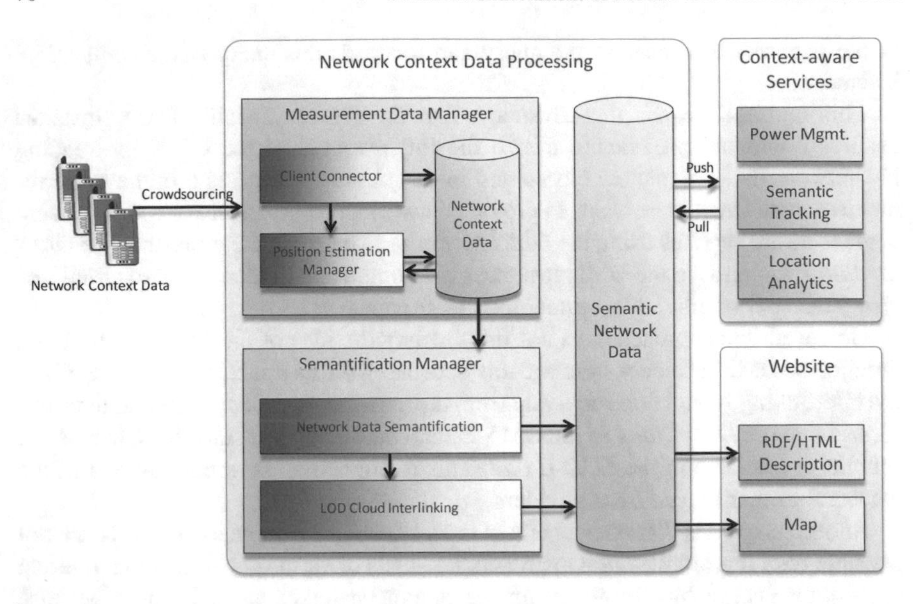

Fig. 4.5 OpenMobileNetwork – functional architecture

the country the measurement came from. In addition, all values are deleted that do not have a valid MNC [72] for the operator, a Cell-ID, and a LAC. After the filtering process, the network context data is standardized to the needed format for further processing and is stored into a relational database.

The filtered and accurate network measurements are then passed to the *Position Estimation Manager (PEM)* component, which provides modularized functions for handling the interaction between different position and coverage area estimation algorithms. For a set of measurements with the same Cell-ID (or WiFi AP), the *Position Estimation* module of the PEM returns the cell (or WiFi AP) with an estimated (or updated if already available) position according to the implemented algorithm, whereas the *Coverage Area Estimation* module approximates the cell's coverage area (see Sect. 4.4). If the estimated cell coverage area is very low, a minimum range of 50 m for both UMTS and GSM cells is used.

The design of the *Measurement Data Manager* ensures easy integration of additional modules and reduces the overload when specific algorithms or further processing of measurements are not needed. It is capable of receiving online as well as offline measurements depending on the Internet connection of the smartphone clients.

Having processed the raw measurements and created a set of calculated cells as well as WiFi APs, the *Measurement Data Manager* calls the *Semantification Manager* that is responsible for "semantifying" the data according to the principles of *Linked Data* and for storing it into the triplestore. In order to improve the flexibility of the system, the *Network Data Semantification* component is separated into different modules, which can be easily extended or reduced dependent on the needed dataset or the data that is available for processing. It is designed to process mobile network cell data, WiFi access point information, user and traffic data as well as service usage information separately. The instance data located in the relational database is imported and transformed into *Linked Data* based on the *OpenMobileNetwork Ontology* (see Sect. 4.5.1) that defines a vocabulary for network topologies. Furthermore, the *LOD Cloud Interlinking* module provides functionality to interlink the *OpenMobileNetwork* with external datasets available in the *LOD Cloud* (e.g., *LinkedGeoData* or *DBpedia* [26]) in order to enrich the networks with additional context information. This function takes the geographic coordinates and the coverage area radius of each single cell in the *OpenMobileNetwork* dataset and queries the endpoint of *LinkedGeoData*, for example, in order to get the points of interest within the coverage area of the cell and create a link from the cell to the POI.

The *OpenMobileNetwork* triplestore provides an endpoint, so that context-aware services, such as the power management or *Semantic Tracking* services, can query the triplestore in order to get the required data. According to the *Linked Data* principles, the resources are also represented via HTML and RDF descriptions on the *OpenMobileNetwork* website, which also provides a visualization on a map showing mobile network cells and WiFi APs, their coverage areas and further detailed information when clicking on them.

4.3 Network Context Data Collection

In Sect. 4.2.1, we addressed several architectural alternatives to provide semantically enriched mobile network data and decided to apply a crowdsourcing approach for a proof of concept with one central mobile network cell and WiFi AP triplestore.

In order to maximize the power of crowdsourcing for our purposes, we applied various facets of it including a systematic warwalking and wardriving method shown in Sect. 4.3.1, a gamification approach described in Sect. 4.3.2 as well as a crowdsourcing background service discussed in Sect. 4.3.3. By doing so, we were able to collect as much data as we can from different users using the corresponding apps of their preferred approach. Furthermore, we were able to analyze which approach works best within our environment. The results of this analysis are presented in [C8] and are not discussed within this thesis.

Taking the context data requirements and classification [C3] presented in Sect. 3.2 as a reference, all crowdsourcing methods that are presented in the following subsections conduct a certain set of relevant network context data. The sets differ in their size depending on the application scenarios the crowdsourcing methods and their corresponding apps are used in. Table 4.3 gives a summary of the collected parameters in the form of a list and maps them to the corresponding crowdsourcing approaches.

Table 4.3 List of collected network context data via crowdsourcing

Network Context Type	Network Context Data	Crowdsourcing Type		
		Wardriving	Gamification	Background Service
Mobile Network (Connected Base Station)	Cell-ID	X	X	X
	LAC	X	X	X
	MCC	X	X	X
	MNC	X	X	X
	Network Access Type	X	X	X
	RSSI	X	X	
	PSC	X	X	
	Operator	X	X	
Mobile Network (Neighboring Base Stations)	Cell-ID	X	X	
	LAC	X	X	
	PSC	X	X	
	Network Access Type	X	X	
	RSSI	X	X	
WiFi Network	BSSI	X	X	X
	SSID	X	X	X
	RSSI	X	X	
Mobile Device	Location (WGS84 Coordinates, Accuracy)	X	X	X
	Time	X	X	X
	Operating System	X	X	
	Device Model	X	X	
	Hardware Name	X	X	
	Manufacturer	X	X	
	SIM Card Information	X	X	
	Roaming State	X	X	
	Traffic Consumption	X		
	Service Usage	X		

4.3.1 *Systematic Warwalking and Wardriving*

When there is the need of acquiring network data in a simple and effective way (especially for a first quick proof of concept), applying warwalking or wardriving is the most straight-forward approach. Using these approaches, a community of users manually start the collection function of the designed data collection app, systematically walk or drive through all streets in a specific area where network measurements are needed, and close the collection process after being done. During the collection phase, the app is running in the background allowing users to use their smartphone as usual. These methods have the advantage that network measurements are conducted with high quality in a short amount of time since they are collected in a controlled environment by users who know the purpose of the collection, have a direct benefit out of it, or are enthusiastically contributing with their data.

For building up our network topology triplestore, warwalking and wardriving were the first and most effective approaches that we have applied. By doing so, we have collected static as well as dynamic network context data in order to enable the modeling and visualization of the current and historic state of the network. Static network context data contains mobile network data of the base station the user is connected to (e.g., Cell-ID, LAC, MCC, MNC, etc.), the neighboring base stations as well as data about nearby WiFi APs (e.g., BSSI, SSID, and RSSI). Dynamic network context data, on the other hand, consists of information about the user's mobile device (e.g., location, time, device model, roaming state, etc.), the traffic consumption generated on the smartphone or information about the services used on it.

The corresponding *OpenMobileNetwork for Android (OMNApp)*[8] app that we have implemented for warwalking and wardriving purposes is described in detail in Sect. 6.2.1.

4.3.2 *Crowdsourcing via Gamification*

Collecting network topology measurements in order to estimate a full mobile network map and integrating dynamic as well as live network context data requires a crowdsourcing approach with a significant number of motivated users willing to spend time conducting those measurements and contribute with their data. This turns out to be very difficult since most of the users do not have any direct benefit from contributing data, which limits the number of them to open data enthusiasts, test users, and people interested in the project. Another problem is that measurements conducted via wardriving or warwalking are mostly concentrated in high-frequented areas like streets or highways, whereas very few or no measurements are available in rural or hardly reachable areas (e.g., while driving). This leads to measurements not

[8] Available at *Google Play*, https://play.google.com/store/apps/details?id=org.openmobilenetwork.app.

being well distributed, which is, however, very important for accurately estimating network topologies.

We have identified two major challenges to tackle when using our warwalking and wardriving method. The first challenge is to attract a wide range of motivated users contributing with their data, so that we can establish a complete map. This can be done by providing frequently changing incentives based on gamification methods when collecting data. Secondly, we need to make sure that measurements are acquired mainly in locations where no or few measurements were conducted before. This will harmonize the distribution of the measurements systematically collected via wardriving, which will then increase the accuracy when estimating mobile network cell and WiFi AP positions.

4.3.2.1 OMN Measurement Framework

For this purpose, we have designed the *OMN Measurement Framework* [C8] that handles these issues and enables the gamification of the network measurement crowdsourcing process by using *Location-based Games (LBGs)*. This framework allows developers to easily create various applications on top of it with the goal of contributing data to the *OpenMobileNetwork*. By doing so, the incentive can be changed frequently (e.g., in the form of different games) keeping users motivated for a longer period of time. It is also capable of guiding users to locations with a lack of network measurements in order to increase the accuracy of network topology estimations calculated out of a collection of distributed data.

Two main modules are provided by the framework that work independently from each other. The first component is called *Location Provider* and is responsible for obtaining *Framework Measurement Locations (FMLs)* from the *OpenMobileNetwork* based on the current position and connected mobile network of the user, whereas the *Measurement Service* collects network topology data and ensures that this data is transmitted to the *OpenMobileNetwork*. Both functions are independent from each other allowing developers who use the *OMN Measurement Framework* to decide which part they want to use. Figure 4.6 represents a sequence diagram illustrating the workflow of the *OMN Measurement Framework* and the interaction with the LBG as well as the *OpenMobileNetwork* platform.

In the first step, the LBG requests a list of geo points from the framework by sending the current position of the user and the number of points it wants to receive. The framework forwards this request to the OMN and asks for a list of FMLs.

An FML describes a circular area with a central point given in WGS84 coordinates as well as a radius given in meters and represents in our context spots where measurements are needed and users should be guided to in order to collect network topology data. These spots can be determined in different kind of ways:

One approach could be by defining these spots as all areas that are in the vicinity of the user's current position and where no measurements have been collected so far or for the mobile network the user is connected to. Another alternative could be to determine the spots out of a list of manually predefined locations that the platform

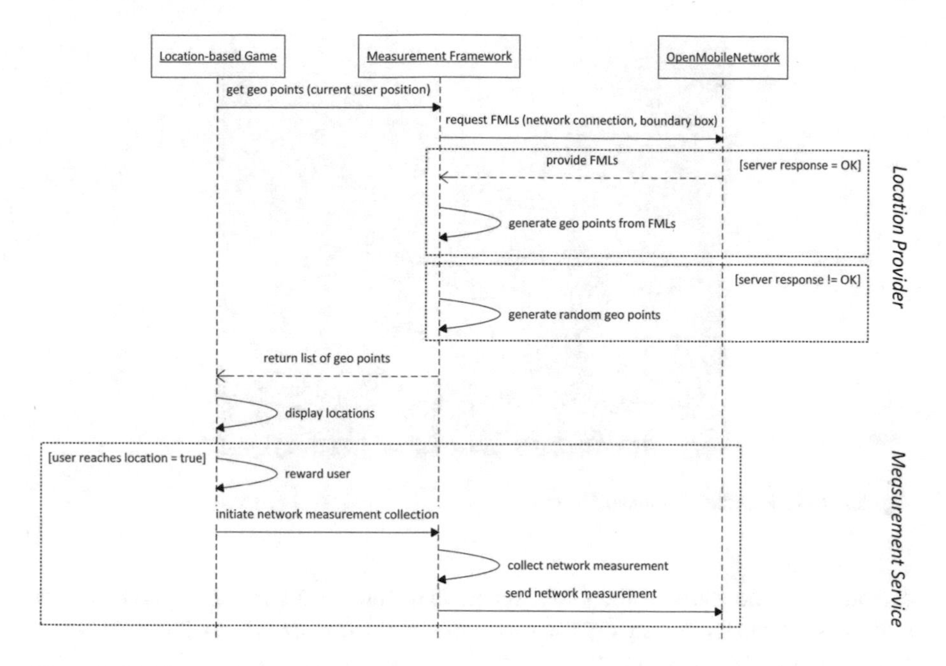

Fig. 4.6 OMN Measurement Framework workflow

owner is interested in to have measurements. The resulting FMLs would have a small radius with the center point being the predefined location of the platform owner. A third possibility could be to define context-dependent spots such as a stadium hosting a certain soccer game where additional base stations are deployed for a temporary period of time. Finally, FMLs could be created in order to update old measurements. In this case, users would be guided to locations where measurements were collected a long time ago, so that the existence of the mobile network cells and WiFi access points at these spots could be verified and updated if needed. As a first proof of concept, we have implemented the first approach and defined spots as areas where no measurements have been collected so far.

In order to obtain FMLs, the framework generates a bounding box based on the location of the user and transmits it to the OMN together with information about the user's network connection. The *FML Provider* function of the OMN creates the requested FMLs based on the current database of measurements and uses the transmitted bounding box in order to ensure that the FMLs are within the reach of the user. For this purpose, the *FML Provider* determines the number of measurements already been conducted within the transmitted boundary box and clusters the existing data within the box in order to generate FMLs outside of the clusters representing locations with few or no measurements (see Fig. 4.7). This enables the framework to send users to locations around high frequented areas even when boundary boxes are created close to those areas. Having a set of calculated FMLs, the server provides these FMLs to the framework out of which a list of geo points are selected and

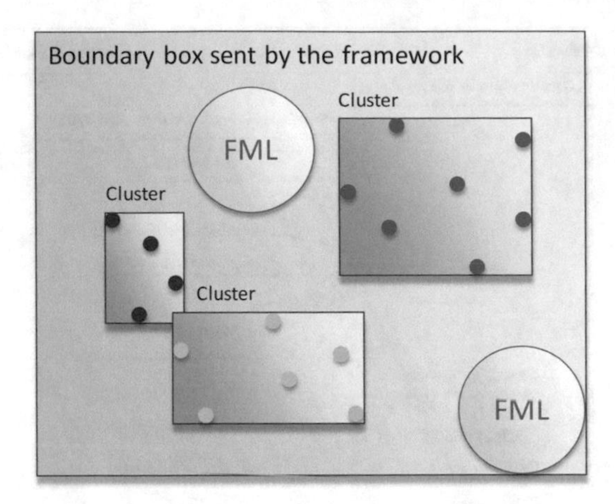

Fig. 4.7 FML Provider – boundary boxes

forwarded to the game. If the OMN server is offline or unreachable, geo positions within the boundary box are created randomly by the framework, so that the delivery of positions to the game is always ensured.

As soon as the user reaches one of the calculated locations, he is rewarded according to the game rules and the network data collection of the *Measurement Service* module is initiated by the LBG. Even though the framework could determine automatically whether the user has reached a relevant location or not (by constantly comparing the user's current position to the target location, for example), we designed the framework more flexible and enabled the game to trigger the network collection, so that different measurement approaches could be applied by potential game developers.

During the measurement process, the current position of the user is determined via GPS. Once the GPS fix is obtained with sufficient accuracy, information about WiFi APs in vicinity as well as mobile network data containing all available information about the network cell the user is currently connected to and all neighboring cells within range (see Table 4.3) is collected. After the collection process, the data is first stored locally in order to ensure that no information gets lost due to network connectivity issues, for example. In the next step, the data is transmitted to the OMN where it is added to the database and further processed for network topology estimations (see Sect. 4.4). Since the database is also used to generate the FMLs, no other user will be sent to the same location again. If the server is not reachable, the local copy of the measurements is kept. Otherwise the local copy is removed and previous measurements, which have not been transmitted, are sent to the server.

4.3.2.2 Location-Based Game Design

Game Patterns. In order to realize an LBG that fully utilizes the potential of the *Measurement Framework*, we analyzed four common game patterns [89] that are used for designing LBGs.

One of the most popular game patterns is *Search-and-Find* in which a player needs to find a certain location given as geo coordinates or concrete real-world places (e.g., POIs). Here, the player is either required to search for the location or he is guided to the place with the ultimate goal of reaching the destination in order to get a reward. A famous game genre using this pattern is *geocaching*[9] where a player needs to find a real-world box hidden at a location given as geo coordinates. This box contains various items out of which the player can take one out and put a new item in return.

In contrast to *Search-and-Find*, the *Follow-the-Path* game pattern does not focus on the destination, but rather on the route the player is taking while moving towards the destination. One variant of this pattern provides a predefined route that a player has to follow with negative effects on the game result if the player changes the route. Other versions do not predefine a route, but monitor the user's movement in order to provide this information to other gamers.

Chase-and-Catch [105], on the other hand, is a pattern used for games with the goal of hunting moving objects. Here, a player can hunt another (moving) player or a virtual object that has a frequently changing location representation in the real world.

Finally, the *Change-of-Distance* pattern represents games in which the target location is irrelevant, but rather the movement towards and away from it plays a significant role. Players are usually rewarded in such games after having moved a certain distance.

Design Decisions. In their paper, Rula et al. [128] point out that controlling the movement of users through gamification (e.g., in order to collect measurements) leads to better results than just giving them the freedom to move to places of their choice and contribute with data. Taking this finding as a reference, we also came to the conclusion that our crowdsourcing user base should be guided to specific locations where network topology data is needed in order to harmonize the distribution of the measurements. Therefore, we analyzed the game patterns mentioned above and determined the ones that fit best to this requirement.

A game based on the *Search-and-Find* pattern, for example, could send users to specific locations where data has not been collected before. This game could be combined with the *Change-of-Distance* pattern, so that users that arrived at the destination could be further asked to move a certain distance in different directions for also measuring the network topology in the whole area around this place. Another alternative could be a *Follow-the-Path* game that could instruct users to use a certain route where network topology data is of high interest. In addition, a much more complex game using the *Chase-and-Catch* pattern could lead users from one location

---

[9]https://www.geocaching.com/.

to the other by moving the target object to be caught to places with lacking network topology data.

Each pattern is more or less applicable for our purposes. However, when considering the trade-off between complexity of the implementation and the effectiveness of the applied game pattern for the purpose of network data collection, we think that *Search-and-Find* as well as *Follow-the-Path* fit best for this task. Therefore, within our proof of concept, we decided to build an LBG based on the *Search-and-Find* game pattern. The implemented *Jewel Chaser*[10] game and its features are demonstrated in detail in Sect. 6.2.2.

4.3.3 Crowdsourcing as a Background Service in another App

As described above, the systematic warwalking and wardriving as well as crowdsourcing via gamification approaches are both applied with the sole purpose of collecting network topology data for the OMN. The users of apps based on these methods are fully aware of their data contributing function and either manually start and stop collecting data or are guided to places where data is needed via a game. However, if users do not use the warwalking and wardriving app or do not play an LBG based on the *OMN Measurement Framework*, no data is collected for the OMN even though they might visit interesting locations in between where network measurements are very relevant. Therefore, it is of great interest that we constantly enrich the OMN dataset in an indirect way by also utilizing apps where the main purpose of them is rather different than collecting network topology data.

We applied this crowdsourcing strategy within our *Context Data Cloud for Android (CDCApp)*[11] [C6] application. The main purpose of the *CDCApp* is to collect personal location preference data in combination with the current contextual situation for contributing to the LCD dataset (see Sect. 5.2 for further details). However, as a positive side effect, mobile network and WiFi AP topology data is also acquired during this process enriching the OMN dataset with new measurements.

The LCD context data collection approach of the *CDCApp* is illustrated in Sect. 5.2.1, whereas app details for conducting network measurements are given in Sect. 6.2.3.

---

[10] Available at *Google Play*, https://play.google.com/store/apps/details?id=org.openmobilenetwork.jewel.

[11] Available at *Google Play*, https://play.google.com/store/apps/details?id=org.contextdatacloud.app.

4.4 Network Topology Estimation

Context-aware services that are based on mobile network topology data (as we exemplary present in Chap. 7) require a very accurate geographic mapping of the network topology in use. In location-based services, for example, the accuracy of the target's position as well as the distance to his contextual environment is a crucial factor influencing the user experience. Depending on the service scenario, this requirement can be met using a coarse-grained (e.g., *Semantic Tracking* for restaurant recommendations in the vicinity) positioning solution that utilizes the mobile network the user is subscribed to. Given that real network topology data is not accessible and we have decided to apply a crowdsourcing approach in Sect. 4.2.1 for identifying network topologies, context-aware services with a high user experience can only be provided if the estimated network topology data is of high quality. Hereby, quality is defined by the distance of an estimated mobile network cell to its real position as well as a high approximation of its coverage compared to the real coverage area.

Since we want to assure high data quality in the *OpenMobileNetwork*, we describe and analyze different algorithmic approaches, which we have considered for estimating the location and coverage area of a mobile network cell. The basis for each algorithm is a set of measurements that belong to a certain cell containing a received signal strength value, a GPS position fix and the cell's unique identifier. We conclude this section by presenting the approach that we have finally applied for our dataset.

4.4.1 Centroid-based Approach

As used by Kim et al. [77] for estimating the positions of WiFi access points, the standard centroid-based approach assumes that the definite position of a (mobile network or WiFi AP) cell is the barycenter of all available positions from the set of measurements for this specific cell. In case of a finite set of points S, Eq. 4.1 defines the centroid of a calculated cell position P as the arithmetic mean, where P_k is the position of the kth measurement and n represents the number of all measurements made for this cell:

$$P = \frac{1}{n}\sum_{k=1}^{n} P_k \tag{4.1}$$

This approach has a complexity of *O(n)* and is used by *OpenCellID*, for example. However, the accuracy of an estimation strongly depends on the distribution and density of the measurements since the position of a cell will always be shifted to areas in which the measurements for this cell are more dense. If users tend to travel on a main street more frequently than on side streets, for example, the calculated position of the cell (covering the main street as well as the side streets) will be more shifted to the main street. This, however, might not indicate better network coverage on the main street or correspond to the real position at all. If the dataset increases with measurements collected via (uncontrolled) crowdsourcing instead of systematic wardriving, the impact of unequally distributed measurements will also increase.

4.4.2 Weighted Centroid-based Approach

In order to improve the accuracy of the standard centroid-based approach, Kim et al. [77] also applied a weighted centroid-based algorithm, which extends the standard centroid-based approach by a weighting factor. They utilized the received signal strength at each measurement position P_k for determining the weight w_k with the assumption that a good signal strength corresponds to a close proximity to the base station and vice versa (see Eq. 4.2).

$$P = \frac{1}{\sum_{k=1}^{n} w_k} \sum_{k=1}^{n} w_k * P_k \tag{4.2}$$

Here, the weight w_k is linearly mapped to a value between 0 and 1 with 0 being assigned to the minimum received signal strength $rssi_{min}$ and 1 corresponding to the maximum received signal strength $rssi_{max}$ of the cell's measurement set S. As stated in Eq. 4.3, the weight w_k of the kth measurement is calculated by using $rssi_{min}$ as well as $rssi_{max}$ and the received signal strength $rssi_k$ of the kth measurement.

$$w_k = \frac{rssi_{\mathrm{k}} - rssi_{\mathrm{min}}}{rssi_{\mathrm{max}} - rssi_{\mathrm{min}}} \tag{4.3}$$

Similar to the standard centroid-based approach, the computational complexity of this algorithm is also $O(n)$. Due to the fact that the weight is determined by the received signal strength, the impact of measurements with a poor signal strength is reduced on the position estimation, which leads to much more accurate results (see Sect. 9.2). However, the assumption that a good signal strength corresponds to a close base station proximity is not always given. If the base station is at the roof top of a building, for example, measurements inside the building will most likely have a bad signal strength (due to directed antennas and shadowing effects) even though the base station is very close. In such cases, this algorithm might also not perform very well.

4.4.3 Grid-based Approach

Nurmi et al. [109] proposed a grid-based algorithm to estimate the position of a mobile device by using a fingerprinting positioning technique. In our work, we have utilized this algorithm for calculating the position of mobile network cells. Similar to the weighted centroid-based approach, we assumed that a good signal strength is an indication for a close proximity to the base station.

In this approach, the geographic map is divided into several grid cells with an edge length of d for each of them. Taking the position of a network measurement as a reference, each measurement is mapped to its corresponding grid cell. Afterwards, a fingerprint is calculated for each grid cell based on the mean μ and the standard deviation σ of all received signal strength values inside of it.

Having the fingerprints, the maximum received signal strength $rssi_{max}$ for the processed mobile network cell j is determined in the measurement set S. In a second step, the probability p of observing the signal strength $rssi_{max}$ is calculated for every grid cell i using a Gaussian distribution (see Eq. 4.4).

$$p(i|j, rssi_{\max}) = \frac{1}{\sqrt{2\pi\sigma_{i,j}^2}} \exp\left(-\frac{(rssi_{\max} - \mu_{i,j})^2}{2\sigma_{i,j}^2}\right) \tag{4.4}$$

The grid cell i with the highest probability p is taken as the location of the mobile network cell. In order to determine the geographic position of the cell, the center of this grid cell is mapped back to a latitude and longitude value.

Due to the fact that measurements with a close proximity are mapped to the same grid cell and an average signal strength value is calculated, the grid-based approach suits very well for overcoming the problem of unequally distributed measurements and delivers very accurate position estimations (as shown in Sect. 9.2). However, even though the complexity of this algorithm is delimited by $O(n)$, the probability calculation and the reverse latitude/longitude mapping are computationally very expensive. Another drawback of this approach is that the position estimation accuracy depends on the granularity of the grid. Nurmi et al. used a granularity of 20 m, which is completely sufficient for our work as we do not expect better accuracy.

4.4.4 Minimum Enclosing Circle

One major disadvantage of the centroid-based approaches described in Sects. 4.4.1 and 4.4.2 is their dependency on the distribution of the measurements since the centroid of a cell is calculated as the arithmetic mean of the measurement positions and is shifted to more dense areas. This problem can be avoided by rather defining an outline shape out of the measurement positions for a specific cell and calculating the position of this cell as the centroid of the defined shape.

The minimum enclosing circle, which we have used within our studies, is one approach for identifying the outline from a set of points. It is defined as the circle with the smallest radius enclosing all points from the set S of measurements. Megiddo [104] first solved the mathematical problem behind it in a linear amount of time $O(n)$

by using pruning techniques. Skyum [142], on the other hand, proposed another algorithm with a complexity of *O(N*log(N))* simplifying the algorithm of Megiddo.

Due to the fact that this approach encloses all measurements for a specific cell inside a circle, it is also suitable for estimating the coverage area of this cell. The basic assumption is that coverage is given everywhere inside the circle even if there are locations where no measurements have been conducted. This approximation provides a minimum range for every (mobile network and WiFi AP) cell and is also applicable if the number of available measurements is low.

However, the circular coverage area estimation might deliver poor results in several cases: If the available measurements are located next to each other in a very dense environment, for example, the calculated minimum enclosing circle might be an underestimation of the actual coverage area. On the other hand, collected measurements with potentially false positions (due to the nature of crowdsourcing) can lead to huge and unrealistic coverage area estimations. Another drawback is that, especially in urban areas, walls, terrain, and other obstacles [17] highly affect the network coverage, so that a circular estimation is not very fitting.

The minimum enclosing circle approach is quite simple to implement, overcomes the problem of the centroid-based approaches and might save computation time if utilized for position as well as coverage area estimation at the same time. However, the number of collected measurements inside the circle including their received signal strength values have no impact on the position estimation at all even though these parameters might give an indication about the strength of network coverage and vicinity to the base station (as described in Sect. 4.4.3).

4.4.5 Signal Maps based on Crowdsourcing

In order to accurately estimate the coverage area of a mobile network cell, Mankowitz and Paverd [97] apply a crowdsourcing approach and collect network measurements with signal strength information via a client software. Having a sufficient set of measurements, a signal map can be created that illustrates the signal strengths of a cell in different spots such as inside or outside a building. According to the authors, this approach gives a very good and realistic understanding of the network coverage boundaries of an operator and is very useful for identifying coverage lacks.

However, due to the nature of crowdsourcing, the signal map strongly depends on the contribution of the user community. If there are areas that are not publicly accesible, for example, coverage information will not be given for these areas. Here, we might need to work with estimations again in order to fill the coverage gap.

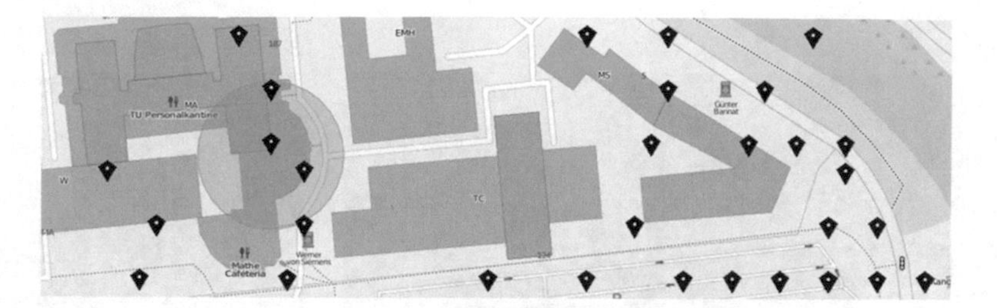

Fig. 4.8 Grid-based position estimation – WiFi access points side-by-side

4.4.6 Applied Topology Estimation within the OpenMobileNetwork

We have applied all position estimation algorithms that we have introduced above to a subset of the *OpenMobileNetwork* dataset and performed a detailed accuracy evaluation on the approximated mobile network cell positions, which is shown in Sect. 9.2.

Taking the achieved evaluation results as a reference, we initially decided to use the grid-based algorithm as our final position estimation approach for the *OpenMobileNetwork*. However, in comparison to other methods, the grid-based approach was computationally very intensive, so that the data processing time from the collection up until to the visualization increased to a great amount. Furthermore, the grid visualization on the map with all cells and WiFi access points being systematically positioned side-by-side and additionally several cells as well as WiFi access points overlapping each other on the exact same position did not look very "realistic", which is exemplary illustrated in Fig. 4.8 for a certain city area in Berlin. Therefore, in the course of time, we switched the position estimation as well as the map visualization to the second-best performing weighted centroid-based approach.

For the coverage area, we applied a hybrid approach consisting of several methods that we discussed above and estimated the coverage area of a cell (or WiFi AP) as a circle as well as a polygon.

Figure 4.9a shows an example for the circular coverage area estimation, which uses the weighted centroid-based approach for calculating the position of the cell, a signal map for identifying the cell coverage boundaries and the minimum enclosing circle for defining the outline shape out of the measurement positions. In contrast to Sect. 4.4.4, the minimum enclosing circle is not used to estimate the position of the cell (based on the outline shape). Instead, the radius between the calculated position of the cell and the measurement with the longest distance to this position is determined and used as the boundary for the coverage. This enables a dynamic recalculation of the coverage area based on new incoming measurements and an increase of its size if needed. The signal strength value that is attached to each cell measurement could also be used to concentrate the coverage only to areas with a high or medium signal strength (based on the signal map). However, due to simplicity reasons, this approach is not implemented in the current version.

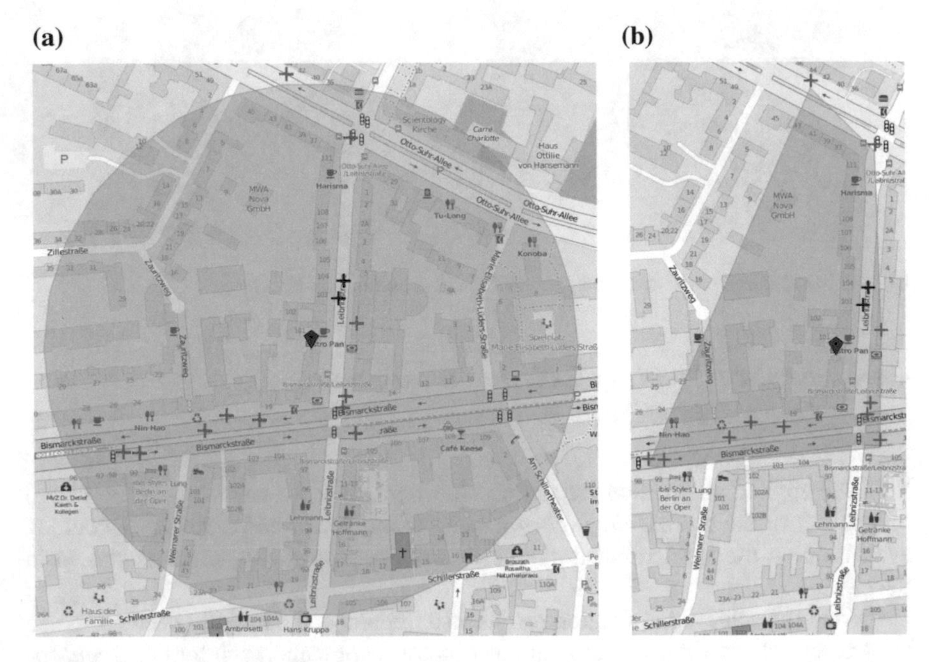

Fig. 4.9 Estimated coverage area shapes within the OpenMobileNetwork **a** Circular coverage **b** Polygonal coverage

When estimating the position and coverage area of a WiFi AP, we are required to perform additional processing steps compared to mobile network cells, because an analysis of our measurements showed that some WiFi APs were measured multiple times in completely different geographic areas causing false position and coverage area estimations. This could have several reasons: First, the location of a WiFi AP could change when people move homes, for example. Furthermore, mobile devices that use the tethering function and are in reach when collecting measurements could also lead to a high amount of different locations for the same access point. Finally, we actually recognized a bug in the WiFi scan function of *Android* (at the time of the implementation) without finding a solution for it. Some WiFi APs were measured multiple times distributed all over a city, for example, even though they did not move at all or were never located in these areas.

In order to avoid having access points in multiple locations, we analyze the number of measurements for an access point in dependency of its location. At first, we take the measurements in which a WiFi AP has been seen for the first time as the position that approximately corresponds to its real location area. We set the maximum size of its coverage radius to 100 m and count measurements within this area as being valid for the access point. If the access point is scanned in another region and the number of measurements exceeds the valid ones collected in the old location, we assume that the access point has moved and hence remove the old location for the access point. The tethering function problem is tackled by completely filtering out obvious mobile SSID names (e.g., *User's iPhone*) or BSSIDs that have a very high number of different location measurements.

Figure 4.9b illustrates an example for a coverage area in a polygon shape. In contrast to the circular coverage area, we use the signal map for defining a polygon that the measurements with the longest distance to the cell's position shape. The location of the cell is not recalculated and is used from the circular coverage area estimation approach.

4.5 Semantification of Network Context Data

The "semantification" of the collected and processed network context data happens on the *Terminological Box (TBox)* as well as *Assertional Box (ABox)* level. Here, the TBox level comprises the modeling of the mobile network and WiFi AP domain including all concepts in the form of the *OpenMobileNetwork Ontology* (see Sect. 4.5.1), whereas the ABox level incorporates the triplification of the concrete instances (see Sect. 4.5.2). Ontology meta data is provided via a VoID description, which is presented in Sect. 4.5.3.

4.5.1 *OpenMobileNetwork Ontology*

The foundation of the *OpenMobileNetwork* is the *OpenMobileNetwork Ontology* (see footnote 13 in Chap. 1) consisting of a set of static and dynamic network context ontology facets that describe mobile networks and WiFi access points from a topological perspective.

In the process of modeling this ontology, we compared a top-down method to a bottom-up approach identifying the advantages and disadvantages of both. In the top-down approach, we analyzed available 3GPP standards [49] and other sources describing mobile network components and their interaction to each other [132]. Based on this analysis, we were theoratically able to create a model mapping all network components in their different generations to an ontology or reuse and extend one of the network ontologies that already exist [40, 120], for example. However, such existing models typically focus on certain aspects of either the network connectivity, mobile devices, or a combination of both, but do not incorporate topological information, i.e., a mapping of (radio access) network components to geographic areas. Moreover, ontologies created via a top-down approach often comprise concepts that are mostly not applicable (or relevant) in concrete application scenarios. In our opinion, a concept for describing channel access methods (e.g., TDMA), for example, does not provide any added value for semantically enriched services.

Therefore, we applied a bottom-up approach where we first analyzed what kind of mobile network data is relevant for implementing semantically enriched context-aware services (described in Chap. 7) and which data we are actually able to collect

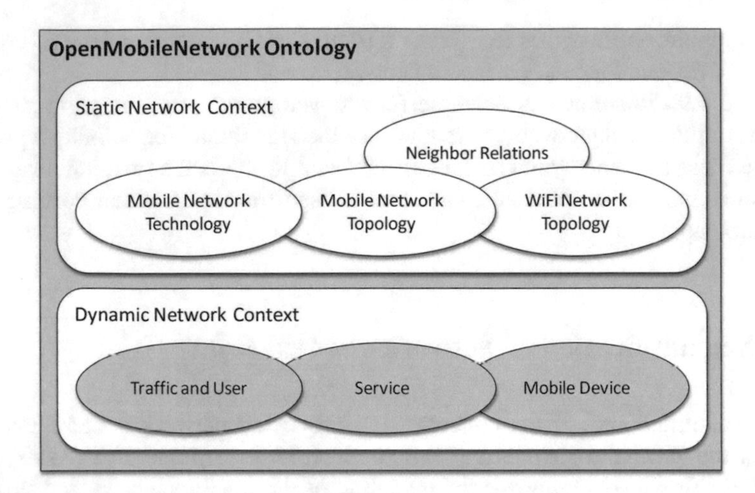

Fig. 4.10 OpenMobileNetwork – network context ontology facets

(e.g., via smartphones) in order to infer the topology of mobile networks and WiFi APs. We took the smartphone capabilities for measuring network characteristics as a reference rather than other ontologies focusing on telecommunication standards. Furthermore, we took privacy issues very seriously. For this purpose, we defined a privacy policy (available online[12] and within the app) stating what kind of data is collected for what kind of purpose. Within this policy, we ensure that no data is collected that can identify a user or a device (e.g., *International Mobile Equipment Identity (IMEI)* or IMSI). Moreover, in the ontology, we only described concepts for aggregated data that do not allow the identification of an individual user such as statistical information for a cell including the number of device models.

Hence, the novelty of the ontology facets employed in the *OpenMobileNetwork* stems from the sound representation of network topology data by providing information about coverage areas, adjacency and overlap of cells while also incorporating WiFi APs. Furthermore, in contrast to other ontologies, we utilize the capabilities of *Linked Data* for interlinking our data with valuable information within the *LOD Cloud*.

In its current state, the ontology is based on RDFS [58] and OWL Lite, which is sufficient with regard to our required expressiveness. By doing so, we ensure optimal performance when querying our *Virtuoso*[13] triplestore. The namespace for ontology concepts is http://www.openmobilenetwork.org/ontology/ with the prefix `omn-owl`. Individuals are identified with http://www.openmobilenetwork.org/resource/ and the prefix `omn`. Within the ontology, we reused vocabulary concepts from well-known ontologies such as the *WGS84 Geo Positioning Vocabulary*, the *OGC GeoSPARQL Vocabulary* [116], or the *GeoNames Ontology*.

[12]http://www.openmobilenetwork.org/measurement/privacypolicy.php.

[13]http://virtuoso.openlinksw.com/.

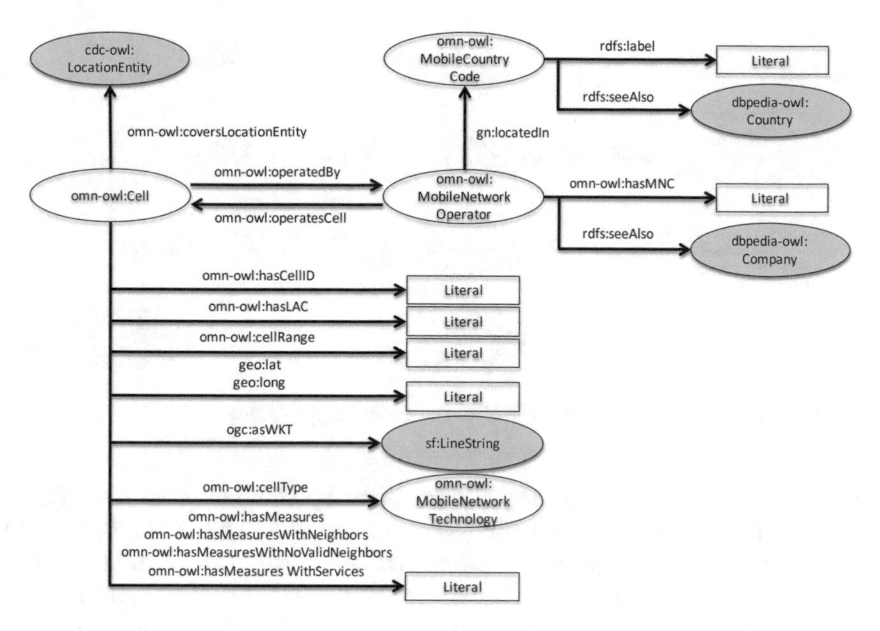

Fig. 4.11 OpenMobileNetwork – Mobile Network Topology Ontology

Figure 4.10 highlights the network context facets of the *OpenMobileNetwork Ontology* by different colors. The ontology facets in white color (*Mobile Network Topology*, *Mobile Network Technology*, *WiFi Network Topology*, and *Neighbor Relations*) represent static network context data, whereas the grey colored ontology facets (*Traffic and User*, *Service* as well as *Mobile Device*) describe dynamic aspects of the mobile network.

4.5.1.1 Static Network Context

The static network context is represented by the *Mobile Network Topology*, *Mobile Network Technology*, *Neighbor Relations*, and *WiFi Network Topology* ontology facets. It refers to network-related information that is expected not to change often or anytime at all. In this context, we can differentiate between *semi-static* and *static* information. An example for static information is the MCC being an identification number for a certain geographic area (e.g., `262` for Germany) [73], whereas a semi-static information could be the location of a WiFi AP that changes whenever the owner moves his home.

The main static concepts of the *OpenMobileNetwork Ontology* are described within the *Mobile Network Topology Ontology* (see Fig. 4.11) providing detailed information about a mobile network cell. With each collected network measurement that is used for estimating a mobile network cell and its characteristics, a list of necessary parameters is recorded, which finds its representation in the ontol-

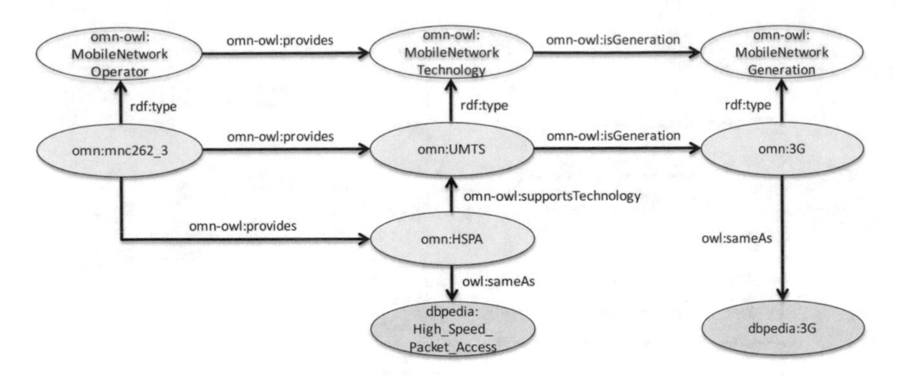

Fig. 4.12 OpenMobileNetwork – Mobile Network Technology Ontology

ogy. An `omn-owl:Cell` is uniquely identified by combining the Cell-ID, the LAC, the MCC, and the MNC to a URI (e.g., `omn:cell34203_20453_262_3`). The `omn-owl:MobileNetworkOperator`, who maintains this cell (described by the inverse properties `omn-owl:operatedBy` and `omn-owl:operatesCell`), is represented by the MNC (`omn-owl:hasMNC`), which in turn is mapped to a string representation of the operator by interlinking it to the corresponding *DBpedia*[14] resource, e.g., `dbpedia:E-Plus` of the type `dbpedia-owl:Company`. This also applies for the MCC (in which the operator is `gn:locatedIn`) since it is also connected to its textual representation within *DBpedia* (`dbpedia-owl:Country`). Furthermore, the cell has its Cell-ID (`omn-owl:hasCellID`), its LAC (`omn-owl:hasLAC`), the network access type (`omn-owl:cellType`), and an estimated position bound to it, which is represented by the `geo:lat` and `geo:long` properties. The calculated coverage area of a mobile network cell is either given as a circle (`omn-owl:cellRange` stands for a radius in meters) or as a polygon defined by the `ogc:asWKT` property using `sf:LineString`. For enabling statistical evaluations, additional meta information is associated to the cell (by using `omn-owl:hasMeasures` and related properties) stating how many measurements have already been conducted.

Using `omn-owl:covers` and its sub-properties (e.g., `omn-owl:coversLocationEntity`), mobile network cells are interlinked with geo-related data from other datasets (e.g., `cdc:loc166` of the type `cdc-owl:LocationEntity`) stating that the related point of interest is in the coverage area of a cell. Here, we leverage the power of *Linked Data* to a great extent. As of September 10th, 2015, this dataset comprised 21,205 cells with 1,442,656 interlinks to points of interest in *LinkedGeoData* and 91,835 relations to location entities in *Linked Crowdsourced Data*. These interlinks are automatically updated whenever the (size of a) cell coverage area changes with new collected measurements.

Figure 4.12 illustrates parts of the *Mobile Network Technology Ontology* including examples on an instance-level (in light brown) that describes what kind of network

[14] *dbpedia-owl*, http://www.dbpedia.org/ontology/.

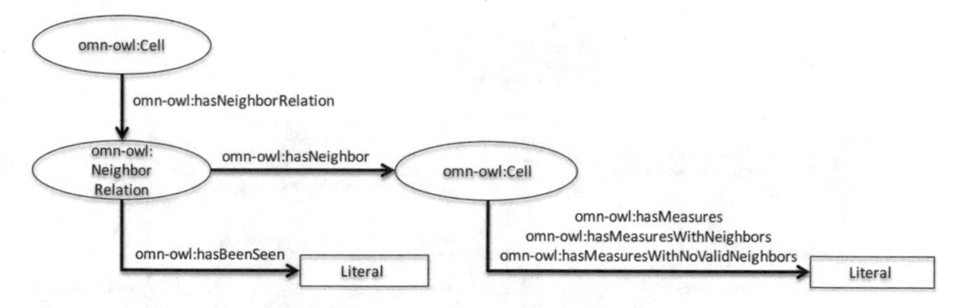

Fig. 4.13 OpenMobileNetwork – Neighbor Relations Ontology

technologies are offered by the network operators (`omn-owl:provides`). It further represents dependencies between different network access types (`omn-owl:supportsTechnology`) and classifies them to mobile network technology generations via the `omn-owl:isGeneration` property. Having this information, an operator (e.g., `omn:mnc262_3`) can categorize his network cells according to network technologies (e.g., `omn:UMTS`) and generations (e.g., `omn:3G`) enabling him to perform a power management in heterogeneous network access technology scenarios, for example.

Every mobile network cell has a set of neighboring cells, which are measured by smartphones contributing to the *OpenMobileNetwork* via *OMNApp*. These cells are expected to be adjacent and thus are viable candidate cells for a handover. The *Neighbor Relations Ontology* (see Fig. 4.13) denotes these neighbor relations using the concepts `omn-owl:NeighborRelation` and the corresponding `omn-owl:hasNeighborRelation`. Two cells are linked to each other via the `omn-owl:hasNeighbor` property. Information about the amount of measurements where a specific neighboring cell has been seen when being connected to a certain cell is provided by the `omn-owl:hasBeenSeen` property, whereas `omn-owl:hasMeasuresWithNeighbors` and `omn-owl:hasMeasuresWithNoValidNeighbors` count the measurements made with or without neighbor relation information to other cells. These counts are used for calculating the probability of having multiple coverage by neighboring cells in that area. This is very crucial in the power management scenario since we need to make sure that coverage is provided by neighboring cells when the serving cell is turned off due to network optimization efforts.

The *WiFi Network Topology Ontology* facet shown in Fig. 4.14 maps the topology of the wireless access points similar to the *Mobile Network Topology Ontology*. An `omn-owl:WiFiAP` is uniquely identified by hashing the combination of the BSSID as well as the SSID and attaching it to the `omn:wifiap` URI (e.g., `omn:wifiap-1823045728`). The BSSID and SSID as well as other parameters, such as the geo position and the circular coverage area information, are also bound to an AP using the `omn-owl:hasBSSID`, `omn-owl:hasSSID`, `geo:lat`, `geo:long`, and `omn-owl:cellRange` properties.

All WiFi APs that are visible when the device is connected to a certain cell are treated as being covered by this cell. This is represented in the ontology by using the

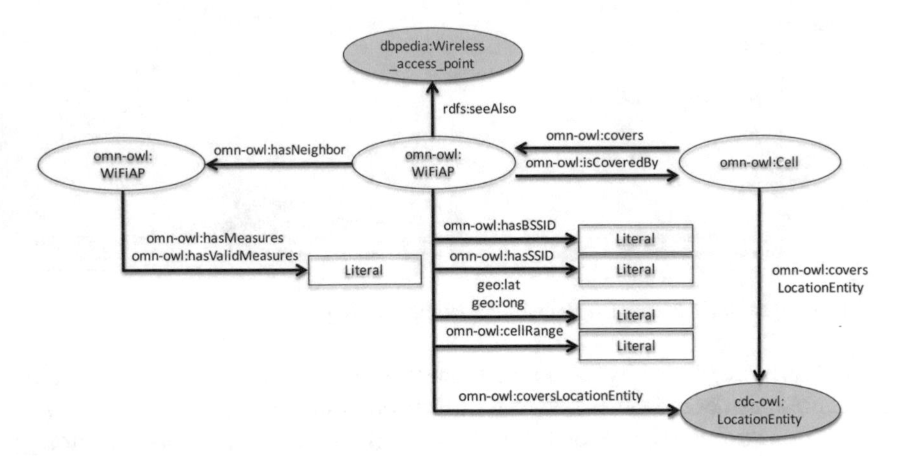

Fig. 4.14 OpenMobileNetwork – WiFi Network Topology Ontology

inverse properties `omn-owl:covers` and `omn-owl:isCoveredBy` for relating a mobile network cell to (multiple) WiFi access points. In addition, WiFi APs are also interconnected with the same geo-related datasets that are used for a linkage with mobile network cells. The geographic mapping of mobile network cells together with WiFi APs and geo-related data can be utilized for providing *Semantic Positioning* solutions (see Sect. 7.2), for example.

As mentioned earlier, we refer to WiFi APs as being semi-static due to the fact that their location can change drastically when people move homes, for example. In order to avoid having access points in multiple locations, we analyze the number of measurements for an access point in dependency of its location. At first, we take the measurements in which a WiFi AP has been seen for the first time as the position that approximately corresponds to its real location area. We set the maximum size of its coverage radius to 100 m and count measurements within this area as being valid for the access point. These numbers are represented by the properties `omn-owl:hasMeasures` and `omn-owl:hasValidMeasures`. If the access point is scanned in another region and the number of measurements exceeds the valid ones collected in the old location, we assume that the access point has moved and hence remove the old location for the access point. A much bigger problem occurs with mobile devices using the tethering function, which leads to a high amount of different locations for the same access point. Here, we try to overcome this problem by completely filtering out obvious mobile SSID names (e.g., *User's iPhone*) or BSSIDs that have a very high number of different location measurements.

Utilizing the calculated locations and coverage areas of the access points, the ontology facet also describes neighbor relations. Two access points are defined as neighbors with the property `omn-owl:hasNeighbor` if the distance between them (calculated from the centroid of each access point's coverage area) is less or equal 150 m. These neighbor relations change dynamically whenever the locations and coverage areas of the WiFi APs change.

4.5.1.2 Dynamic Network Context

Dynamic network context is described by the *Traffic and User*, *Service* as well as *Mobile Device* ontology facets. It refers to data that changes constantly such as the current traffic or the number of users in a mobile network cell. In contrast to static data, dynamic network context enables (real-time) data analytics.

In a power management scenario, knowledge about the amount of traffic generated in a mobile network cell and the number of users connected to it, is fundamental when identifying candidate cells to be de- or reactivated. With each network measurement done by *OMNApp*, we also collect the total incoming as well as outgoing traffic of the smartphone. Applying the concepts of the *Traffic and User Ontology* (as seen in Fig. 4.15) to the collected traffic data, we enable the creation of various traffic and user activity profiles.

An `omn-owl:Cell` has `omn-owl:AccumulatedTraffic` information, which is divided into several `omn-owl:TrafficEvent` instances storing incoming and outgoing traffic data (`omn-owl:hasRxTrafficPerMin` and `omn-owl:hasTxTrafficPerMin`) for every past minute in conjunction with the corresponding timestamp (`omn-owl:hasUniqueTime`). The `omn-owl:latestTimestamp` property, on the other hand, adds the time of the latest traffic data to the `omn-owl:AccumulatedTraffic` information and is updated with every incoming measurement containing traffic information.

User data is stored in a similar fashion. An `omn-owl:UserHistory` is related to an `omn-owl:Cell` and consists of several `omn-owl:UserEvents`. These entities store the number of users for every last minute (`omn-owl:noUserPerMin`) together with the corresponding timestamp (`omn-owl:hasUniqueTime`).

Due to the short time frame of a minute, the `omn-owl:TrafficEvent` and `omn-owl:UserEvent` instances of the *Traffic and User Ontology* support traffic as well as user queries for specific time ranges and enable live, daily as well as historic

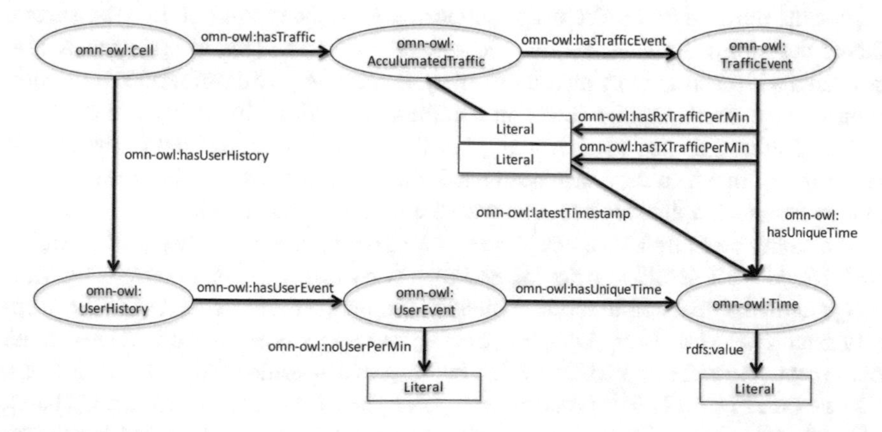

Fig. 4.15 OpenMobileNetwork – Traffic and User Ontology

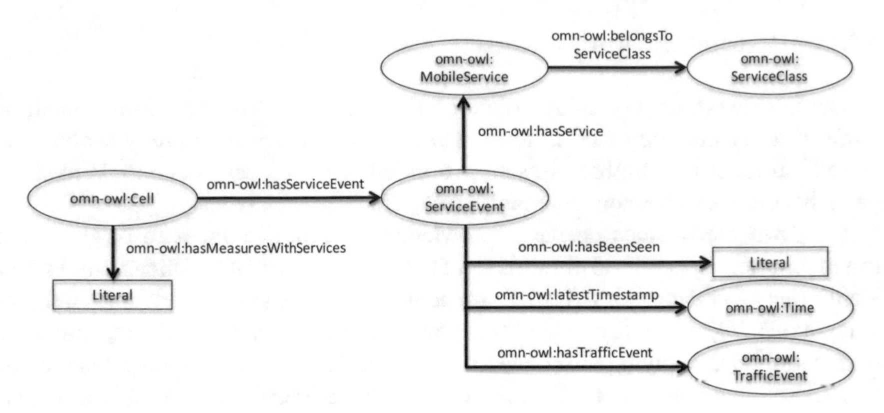

Fig. 4.16 OpenMobileNetwork – Service Ontology

traffic and user activity profiles with high granularity. In addition, this ontology allows prediction algorithms to utilize this data for forecasts.

Information about the service usage behavior of users connected to the mobile network in combination with the capabilities of their mobile devices is another essential piece of network context data that can be utilized for innovative context-aware services. For example, in order to guarantee high QoE when optimizing the network for power management purposes, the operator needs to make sure that the users are able to use any mobile service without limitations in capacity. This can only be provided if the operator is aware of the services being used (frequently and mostly) in a mobile network cell and what kind of technological capabilities the devices of the users have, so that they can be handed over to cells that fulfill the service requirements when deactivating mobile network cells. Another usage scenario can involve mobile advertisement providers that can adjust their location-based advertisement channels based on the services mostly used in a certain region.

For the purpose of service usage support, we extended our ontology by several facets describing services, service classes, and devices. This information is also collected with each network measurement by aggregating the list of services currently running on a user's mobile device in combination with its incoming and outgoing traffic. Through the `measurement_id`, these services are related to the mobile network cell in which they have been used. Figure 4.16 shows the *Service Ontology* that represents mobile services used within a mobile network cell.

An `omn-owl:Cell` has an `omn-owl:ServiceEvent` (e.g., `omn:cell34203_20453_262_3_org_openmobilenetwork_app`) comprising various parameters of a mobile service collected within the past minute. Using the property `omn-owl:hasService`, this `omn-owl:ServiceEvent` is related to an `omn-owl:MobileService`, which belongs to a specific `omn-owl:ServiceClass` (see Fig. 4.17). The property `omn-owl:hasBeenSeen`, on the other hand, indicates how often the respective `omn-owl:MobileService` has been seen within the `omn-owl:ServiceEvent` when collecting measurements.

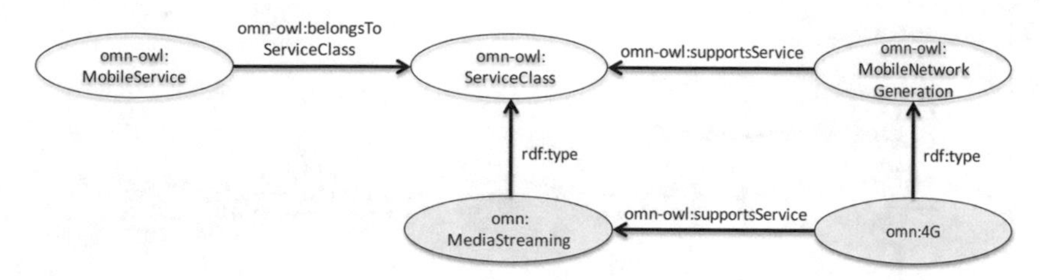

Fig. 4.17 OpenMobileNetwork – Service Classification Ontology

Moreover, each `omn-owl:ServiceEvent` has an associated `omn-owl:TrafficEvent` in combination with a corresponding timestamp (`omn-owl:Time`) showing how much traffic this service has generated for the same minute interval. By relating the amount of traffic produced by a single service to the total traffic in the cell modeled in the *Traffic and User Ontology*, "popular" services used in a cell can be determined. This information is valuable with respect to the aforementioned power management scenario, because if a traffic-intense service, such as a video streaming service, is popular, shutting down this specific cell might not be feasible without negatively affecting the QoE of the connected users.

The ontology facet shown in Fig. 4.17 introduces exemplary parts of the employed *Service Classification Ontology*. This model includes manifold service classes (e.g., `omn-owl:MediaStreaming`) that are based on QoS parameters such as transmission speed or delay. Based on this service classification, we can categorize mobile services into classes with similar requirements using the property `omn-owl:belongsToServiceClass` and also relate them to `omn-owl:MobileNetworkGenerations` (via the property `omn-owl:supportsService`) that are able to deliver the service with the required capabilities and optimal QoE for the user.

Currently, the classification of services is done manually since we did not find an optimal solution for automatically detecting the service classes of processes running on a smartphone. *Android* deprecated the possibility to scan the ports the services are running on (e.g., TCP/UDP) after API Level 18. *Kismet Wireless* provides the *Android PCAP*[15] utility for capturing raw 802.11 frames. This tool, however, needs additional hardware equipment and at least *Android 3.2*. Due to the fact that the *OpenMobileNetwork* is based on a crowdsourcing approach, we cannot expect users to plug-in additional equipment or to keep their operating system with a certain version in order to contribute with service usage data.

In addition, we analyzed the categories selected for an *Android* app in the *Google Play Store* for deriving the service class out of those categories. However, these categories are often too imprecise, broad and arbitrarily selected by the app developer, so that a clear classification according to the QoS classes is also not possible.

[15]http://www.kismetwireless.net/android-pcap/.

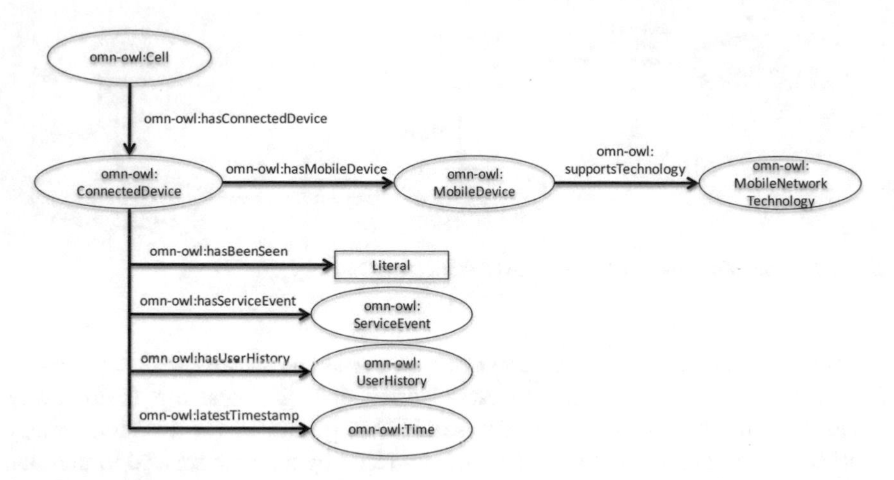

Fig. 4.18 OpenMobileNetwork – Mobile Device Ontology

Figure 4.18 exemplifies parts of the *Mobile Device Ontology* that semantically describes mobile devices connected to the mobile network. Information about mobile devices (e.g., operating system, brand, device name, hardware name, manufacturer, model name, product name, and SIM card information) is also collected during the crowdsourcing process and is further related to the network access type these devices were connected to when performing measurements. By doing so, we can derive information about the number and type of devices currently as well as historically connected to a certain cell, which enables the operator to reconfigure the network based on the capabilities of the remaining user's devices.

As soon as a network measurement with device information is processed, an instance of the `omn-owl:ConnectedDevice` concept with a corresponding `omn-owl:latestTimestamp` is related to the `omn-owl:Cell` the measurement has been done for using the `omn-owl:hasConnectedDevice` property. This `omn-owl:ConnectedDevice` represents a certain `omn-owl:MobileDevice` and supports various `omn-owl:MobileNetworkTechnologys` (e.g., `omn:HSPA`). Additional information to the model of the `omn-owl:MobileDevice` (e.g., `omn:GT-I9100`) as well as the various network technologies is acquired through interlinks to *DBpedia* (e.g., `dbpedia:Samsung_Galaxy_S_II`). The property `omn-owl:hasBeenSeen`, on the other hand, counts how often this mobile device has been measured while being connected to this `omn-owl:Cell`.

Via the `omn-owl:ServiceEvent` and `omn-owl:UserHistory` concepts, the *Mobile Device Ontology* is also interlinked to the *Service* as well as *Traffic and User Ontology* enabling the correlation of service usage and user activity information not only on a cell-level, but also on a device-level.

The *OpenMobileNetwork Ontology* with its different network context facets enables context-aware services with sophisticated semantic queries based on semantically enriched mobile and WiFi network data. Taking the power management scenario as an example, the information required for a deactivation of a candidate cell and the successive handover of the remaining users is captured within the network context ontology facets comprising data about neighboring cells, the connection types supported by the connected devices and the services currently running on these devices.

4.5.2 *Instance Data Triplification*

Taking the defined ontology as a reference, we triplified the approximated network topology data into RDF instances and made the complete dataset accessible through a SPARQL endpoint (see footnote 1 in this chapter). Here, we strictly followed the best practices of publishing *Linked Data* [24] for naming the resources and making them dereferenceable in various formats. We created modular mapping schemas for each network context ontology facet enabling a flexibilty to extend the dataset as well as the *OpenMobileNetwork Ontology* in the future. Please have a look at Sects. 6.3.2 and 6.3.3.1 for further details.

4.5.3 *OpenMobileNetwork VoID Description*

VoID [162] is a vocabulary for describing RDF datasets on a meta level offering end users of linked datasets knowledge about its content. It is currently "submitted for consideration to the W3C's SWIG to publish it as a *W3C Interest Group Note*" and inofficially represents the standard for enriching linked datasets with meta data.

In order to ease the usage of the network context dataset for external users, we have published meta data about the *OpenMobileNetwork* according to VoID, which is available at http://www.openmobilenetwork.org/resource/void.

Figure 4.19 illustrates the main meta concepts of the *OMN VoID Description* that includes information about the dataset contributors (`dcterms:contributor`), the URI defined for the *OpenMobileNetwork Ontology* with the `void:vocabulary` property, the SPARQL endpoint represented by `void:sparqlEndpoint`, the dataset license (`dcterms:license`), the number of triples (`void:triples`), classes (`void:classes`), and properties (`void:properties`).

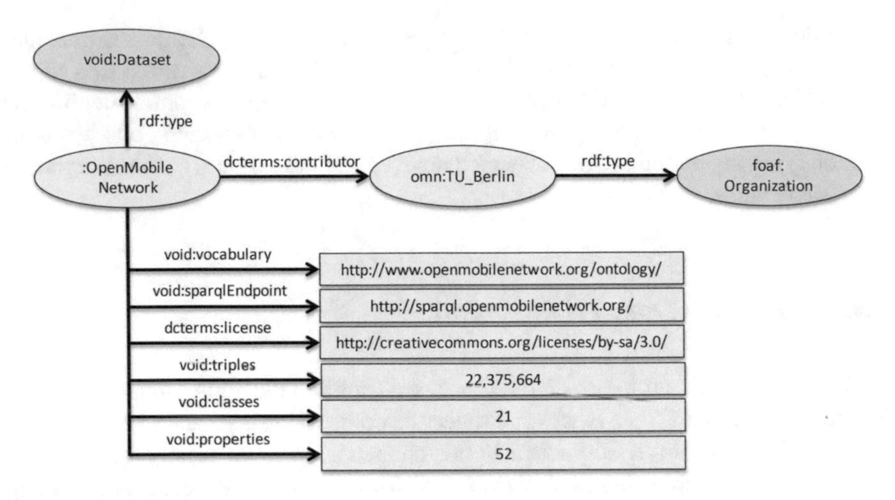

Fig. 4.19 OMN VoID description – information about vocabulary and sample resources

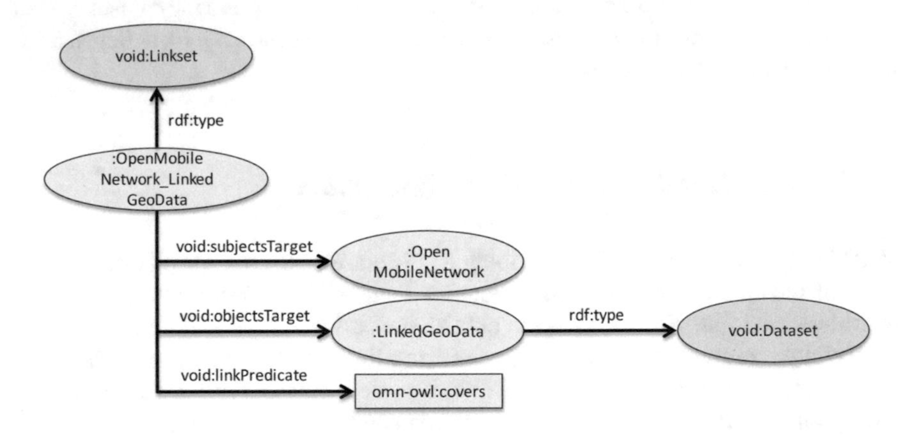

Fig. 4.20 OMN VoID description – information about dataset links

VoID also provides concepts for representing links between datasets within the *LOD Cloud*. Figure 4.20 shows an exemplary `void:Linkset` of the *OpenMobileNetwork VoID Description* (e.g., `:OpenMobileNetwork_LinkedGeoData`) that descibes the interlinking between the *OpenMobileNetwork* and *LinkedGeoData*. The instances in OMN serve as the subject of a triple (`void:subjectsTarget`), whereas the entities in LGD are the objects (`void:objectsTarget`). `omn-owl:covers`, on the other hand, is used as the `void:linkPredicate` to link both datasets.

Chapter 5
Interlinking Diverse Context Sources with Network Topology Data

Chapter 5 highlights the approach of interlinking diverse context sources to the semantically enriched network topology data of the *OpenMobileNetwork* in order to enable sophisticated context-aware services as shown in Chap. 7. For this purpose, Sect. 5.1 describes the interlinking process with already available context data in the *LOD Cloud*. Taking the limitations of the available geo-related linked datasets as a reference, Sect. 5.2 presents *Linked Crowdsourced Data* as a new dataset incorporating not only static, but also diverse and dynamic context information associated with locations. In addition, the *OpenMobileNetwork Geocoding Dataset* is introduced in Sect. 5.3 as another context source that interconnects address data to the *OpenMobileNetwork* for enabling address-related services such as geocoding.

5.1 Interlinking with Available Context Data

In Sect. 3.2.1, we performed a deep dive into the *LOD Cloud* and analyzed what kind of context data sources are available. After designing the *OpenMobileNetwork Ontology* and laying the foundation for a semantic mobile and WiFi network dataset, we selected external datasets based on the *LOD Cloud* analysis that would provide an added value to the OMN and identified two types of data sources to create links to. The first and most important type comprises geo-related datasets providing information about POIs or events, for example, whereas the second type of data sources are rather enriching the network topology data with other semantic representations.

As a first approach, we decided to interlink the OMN to the two most famous datasets in the *LOD Cloud* comprising both types of data sources. We utilized *LinkedGeoData* as a context source for interconnecting geo-related data to the OMN, whereas descriptions in *DBpedia* are used to relate various (textual) representations for mobile network information such as operator names.

A. Uzun, *Semantic Modeling and Enrichment of Mobile and WiFi Network Data*, T-Labs Series in Telecommunication Services,
https://doi.org/10.1007/978-3-319-90769-7_5

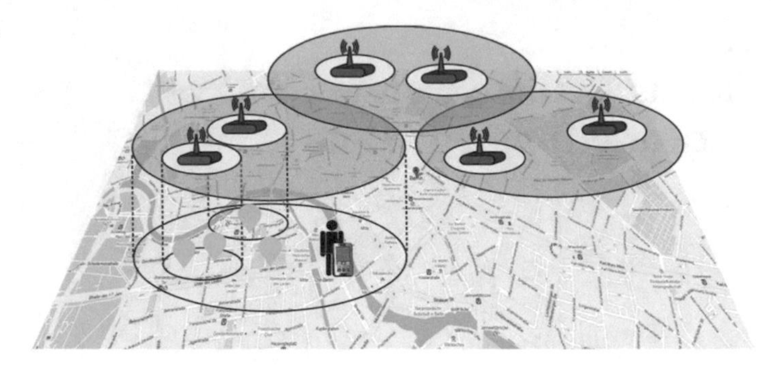

Fig. 5.1 Geographic mapping model of the OpenMobileNetwork

5.1.1 LinkedGeoData

Interlinking geo-related datasets to the *OpenMobileNetwork* is enabled by the *geographic mapping model* that semantically relates (radio access) network components to geographic areas from a topological perspective. As seen in Fig. 5.1, a mobile network cell, which is defined by a position given in WGS84 coordinates and a coverage area shaped as a circle or polygon, covers a certain geographic region. This region comprises a set of WiFi APs with a much smaller circular coverage range. In addition, location-related information, such as points of interest, events, or streets, are also part of the same coverage area. Furthermore, the same geographic area can be covered by multiple mobile network cells (e.g., belonging to other network access technologies or operators) or they can overlap each other to a certain extent.

The geographic mapping model of the *OpenMobileNetwork* semantically describes this "coverage relationship" from a top-down perspective using `omn-owl:covers` and its sub-properties `omn-owl:coversCell`, `omn-owl:coversWiFiAP`, `omn-owl:coversLocationEntity`, `omn-owl:coversPOI`, and `omn-owl:coversAddress`. By doing so, it enables the implementation of a variety of context-aware services such as a power management in mobile networks or a novel *Semantic Tracking* approach for proactive location-based services [C9, C10].

`omn-owl:covers` is a general object property and denotes whether an entity is covered by a cell or WiFi AP. The sub-property `omn-owl:coversCell` is used to create a relation between cells that cover the same area, whereas `omn-owl:coversWiFiAP` is a predicate for describing a WiFi AP that is covered by a cell. Location entities of LCD (see Sect. 5.2) are interlinked with cells and WiFi APs through the `omn-owl:coversLocationEntity` property. `omn-owl:coversPOI`, on the other hand, is used for relating structured points of interest (from geo-related datasets) to radio access network components, while addresses are geographically mapped to cells and WiFi APs using the predicate `omn-owl:coversAddress`. Please note that by the time of development, only `omn-owl:covers` and `omn-owl:coversLocationEntity` were used and implemented for interlinking data to the *OpenMobileNetwork*. Therefore, we will only use these two properties for further explanations.

As described in Sect. 4.5.1, a mobile network cell has an estimated position represented by the `geo:lat` and `geo:long` properties, as well as a calculated coverage area either given as a circle (`omn-owl:cellRange` stands for a radius in meters) or as a polygon defined by the `ogc:asWKT` property using `sf:LineString`. It is put into relation with WiFi APs through the `omn-owl:covers` property describing that an access point is in the coverage area of this cell. The same property is also used in order to interlink geo-related datasets with the *OpenMobileNetwork* stating that a POI is in the coverage area of a cell or WiFi AP. By the time of writing this thesis, the dataset comprised 21,205 cells with 1,442,656 interlinks to points of interest in *LinkedGeoData*. These interlinks are automatically updated whenever the (size of a) cell coverage area changes with new collected measurements.

Links to other datasets are generated by the *LOD Cloud Interlinking* function of the *OpenMobileNetwork* (as presented in Sect. 4.2.2). The *POI-Getter*, which is a specific component of this function, takes the geographic coordinates and coverage range of each single cell in the OMN dataset and queries the triplestore of LGD in order to get the points of interest within the coverage area of the respective cell. The resulting list of POIs is then added to the specific cell via the `omn-owl:covers` predicate.

Listings 5.1 and 5.2 show a SPARQL query example executed by the *POI-Getter* taking a certain cell as an example. First, the geo coordinates and the coverage area for the exemplary cell `omn:cell2160591_5275_262_1` is retrieved through the OMN endpoint. The result of this query is `[52.513382, 13.322177]` as latitude and longitude values as well as 300m as the coverage range of the cell.

```
SELECT ?lat, ?lng, ?cellrange
{
  omn:cell2160591_5275_262_1 geo:lat ?lat .
  omn:cell2160591_5275_262_1 geo:long ?lng .
  omn:cell2160591_5275_262_1 omn-owl:cellRange ?cellrange
}
```

Listing 5.1 Querying *sparql.openmobilenetwork.org* for the geographic coordinates and coverage area radius of `omn:cell2160591_5275_262_1`

In a second step, all nodes being amenities are requested from *LinkedGeoData* that are located within the coverage area of the cell for the given geo coordinates. The result of this query is a list of nodes that represent points of interest within the coverage area of the cell `omn:cell2160591_5275_262_1`.

```
PREFIX lgdo: <http://linkedgeodata.org/ontology/>
PREFIX lgdm: <http://linkedgeodata.org/meta/>
PREFIX geom: <http://geovocab.org/geometry#>
PREFIX ogc: <http://www.opengis.net/ont/geosparql#>
SELECT ?poi
{
  ?poi rdf:type lgdo:Amenity .
  ?poi rdf:type lgdm:Node .
  ?poi geom:geometry [ ogc:asWKT ?geo ] .

FILTER (bif:st_intersects (?geo, bif:st_point (13.322177,
     52.513382), 0.3))
}
```

Listing 5.2 Querying *linkedgeodata.org/sparql* for nodes that are amenities and are in the coverage area of `omn:cell2160591_5275_262_1`

For cells with a very low coverage area approximation (due to missing measurements), we defined a minimum range of 50 m (see Sect. 4.2.2). This is a relatively small radius for a mobile network cell since there are no cells with such a small coverage area in reality (except for home use femto cells). Nevertheless, we used this range also for these cells in the interlinking process in order to ensure that we only relate those points of interest to a cell that are within a close vicinity of it.

Furthermore, in the current version of the *OpenMobileNetwork*, we only considered `lgdm:Nodes`[1] from LGD as links since `lgdm:Ways`[2] are much more complex to map onto cells. However, this is future work to be done and a potential extension of the context sources to be linked.

As mentioned earlier, by the time of development, we used `omn-owl:covers` for interlinking points of interest as well as WiFI APs to cells. However, we realized that we slowed down the performance of SPARQL queries by doing so. For example, if there are services requesting only WiFi APs that are covered by a cell, such as in the *Semantic Tracking* scenario, we still need to query all objects (including WiFi APs and POIs) that are related to the `omn-owl:covers` predicate. Here, we have two alternatives for filtering the WiFi APs. First, we apply a `REGEX` command afterwards and filter all objects that start with `omn:wifiap`. This, however, slows down the performance to a great extent. Second, we extend the query by a statement requesting also a certain feature of a WiFi AP such as its BSSID (see Listing 5.3 as an example). By doing so, we still go through all objects in the first place, but optimize the performance in comparison to a `REGEX` command.

[1] http://wiki.openstreetmap.org/wiki/Node.

[2] http://wiki.openstreetmap.org/wiki/Way.

```
SELECT ?wifiap ?bssid
{
  omn:cell2160591_5275_262_1 omn-owl:covers ?wifiap.
  ?wifiap omn-owl:hasBSSID ?bssid
}
```

Listing 5.3 Querying *sparql.openmobilenetwork.org* for only WiFi APs covered by `omn:cell2160591_5275_262_1`

Nevertheless, the optimal alternative is to directly define sub-properties of `omn-owl:covers` for each type of entity to be related, which we have already done in the *OMN Ontology*. An update of the platform, however, is part of future work.

5.1.2 DBpedia

The dataset of *DBpedia* is mainly utilized in order to enrich the available network information resources with other textual or semantic representations using standard concepts like `rdfs:seeAlso` or `owl:sameAs`. An example for such an enrichment is given within the *Mobile Network Topology Ontology* facet of the *OMN Ontology* (see Fig. 4.11).

Taking the lists of the *International Telecommunication Union (ITU)* for MNCs [73] and MCCs [72] as a reference, we mapped an `omn-owl:MobileNetwork Operator`, who is represented by the MNC (`omn-owl:hasMNC`), to a string representation of the operator within *DBpedia*, e.g., `dbpedia:E-Plus` of the type `dbpedia-owl:Company`. This was also done for the MCC (in which the operator is `gn:locatedIn`) by relating the country code to its textual representation (`dbpedia-owl:Country`).

In addition, the interlinkage with *DBpedia* also allows to crawl for more insights that are related to network information. For example, a link between a network technology and the same resource in *DBpedia*, e.g., `omn:HSPA owl:sameAs dbpedia:High_Speed_Packet_Access` (as done in the *Mobile Network Technology Ontology* facet, see Fig. 4.12), makes additional information, such as an overview about HSPA, available for users of a context-aware service. Another example is given in the *Mobile Device Ontology* facet (see Fig. 4.18), in which additional information to the model of an `omn-owl:MobileDevice` (e.g., `omn:GT-I9100`) is acquired through interlinks to *DBpedia* (e.g., `dbpedia:Samsung_Galaxy_S_II`).

The interlinking is performed by the *DBpedia-Linker* function of the *LOD Cloud Interlinking* component. For this purpose, we assembled a script in which we partially "hard-coded" the links to be created on a concept-level. This was necessary due to the very specific representations of the resources (on both sides) that did not allow for patterns to be identified and automated.

A future outlook in utilizing *DBpedia* could be to interlink district, city, and country names with the network cells in order to optimize SPARQL requests by avoiding the search for locations with the `bif:st_distance` functions.

5.2 Linked Crowdsourced Data

In Sect. 5.1.1, we have described our approach of interlinking the coverage areas of mobile network cells and WiFi APs with geo-related datasets that are already available in the *LOD Cloud*. However, the variety of realizable context-aware service scenarios, such as location analytics, is still limited since geo-related datasets in the *LOD Cloud* are rather of static nature and mainly consist of information such as a name, geo coordinates, an address, or opening hours. Furthermore, even though network providers are able to infer highly frequented locations based on user movements, the results are mostly given on an abstract geographic level without having knowledge about the exact shop on a strip mall, which is highly visited under specific contextual circumstances, for example. Therefore, a linked dataset providing dynamic information about the "popularity" of certain places or the "visiting frequency" of users in specific contextual situations, which is again linked to network topology data, would be of great use for the network operator as well as the LOD community in general.

For this purpose, we present *Linked Crowdsourced Data* [C6] in this section, which is a dataset for location-related data incorporating real user location preferences as well as additional domain-specific information collected via crowdsourcing. The unique selling point of LCD in contrast to other available geo-related datasets is that it links dynamic parameters (e.g., check-ins, ratings, or comments) as well as specific context situations (e.g., weather conditions, holiday information, or measured networks) to the static location data. Moreover, additional domain-specific information that is related to a certain location of interest (e.g., dishes of a restaurant) is also collected via crowdsourcing and connected to this location. This variety of raw data attached to a location in combination with mobile and WiFi network data enables fine-grained and semantically enriched location analytics within the *LOD Cloud* (see Sect. 7.3), which was not possible before.

The dataset is published as *Linked Data* in RDF format through our SPARQL endpoint[3] and a visualization of the data is provided on our *Location Analytics Map* (see Sect. 8.4). With the update of the diagram in January 2017 (see Fig. 5.2), *Linked Crowdsourced Data* officially became part of the *LOD Cloud*.

[3]http://sparql.contextdatacloud.org/.

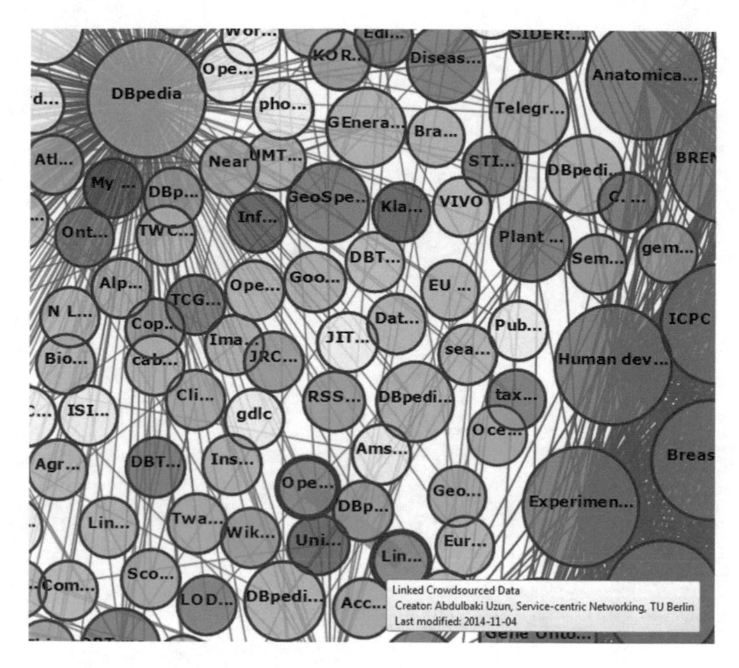

Fig. 5.2 Linked Crowdsourced Data in the LOD Cloud

5.2.1 *Crowdsourced Context Data Collection*

The crowdsourcing community can contribute to this dataset by using the *Context Data Cloud for Android* (see footnote 11 in Chap. 4) application, which is a location-based community app for collecting place preferences of users as well as additional location-related information in combination with their contextual situation. It offers a set of semantically enriched services to users (see Sect. 8.2) and further enables them to check-in to certain POIs in their vicinity providing information about their location visits to other users within their social community. For the data collection and processing part, we have defined a privacy policy,[4] which is approved by the privacy manager of the *Technische Universität Berlin*.

By clicking on the *Local Area* tab within the app (see Fig. 5.3), all location entities within a 500 m radius of the user's current position are retrieved and listed according to their distance. During this request, available information from LCD is queried and enriched with additional POIs from *OpenStreetMap*. Here, we have chosen OSM instead of *LinkedGeoData* due to its permanent availability and up-to-date location entity data. Whenever a user checks in to a certain place for the first time, adds a new location entity (that is not in the displayed list) or modifies existing ones, a `cdc-owl:LocationEntity` instance including its static location information is created or updated in the LCD dataset. In addition, after a check-in to a certain place,

[4]http://www.contextdatacloud.org/index.php?page=privacypolicy.

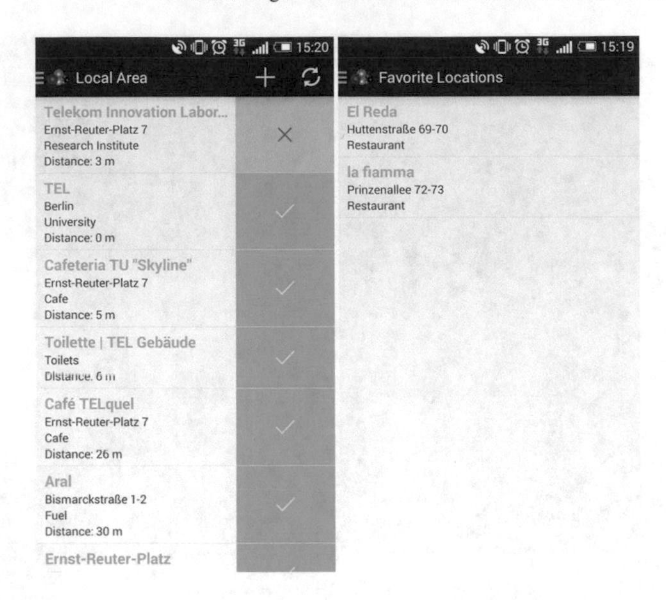

Fig. 5.3 CDCApp – local area and favorite locations

a `cdc-owl:CheckIn` resource is created that `cdc-owl:refersTo` a certain `cdc-owl:LocationEntity`.

Each `cdc-owl:CheckIn` is automatically accompanied by a list of raw context data that is retrieved from Web APIs with *Creative Commons (CC BY-SA)* licenses or by exploiting the sensors on the smartphone. This list includes

- the check-in and check-out time given in minutes after midnight (calculated from the server time),
- the weather conditions by the time of the check-in (retrieved from *OpenWeatherMap*[5]),
- whether the check-in occurred during a regional holiday (retrieved from the *Enrico Service*[6]),
- the information whether the visit happened in an indoor or outdoor environment (using the *IODetector* service [163]),
- the list of WiFi access points that were measured in this area (retrieved from the *Android SDK*), and
- the mobile network cell the user was connected to and its neighbors (retrieved from the *Android SDK*).

The set of attached raw context data is not considered complete and can be extended by other (free) data sources if needed. However, adding new data sources require a description of them within the *Context Situation* ontology facet of the *Context Data Cloud Ontology* (see Sect. 5.2.3).

[5]http://www.openweathermap.org/.

[6]http://kayaposoft.com/enrico/.

5.2.2 *Crowdsourced Context Data Processing*

The LCD dataset is created out of the set of collected raw data by applying a multi-step processing approach for generating higher-level information. After performing a pre-processing of the acquired data on the client side, it is saved in a database and further post-processed as well as cleansed before semantifying and storing it into a triplestore. With each check-in or change of the data on the client side, the LCD dataset is updated on the fly.

Due to the fact that the dataset of LCD is mainly based on OSM, our location entities are categorized using the categories for POIs defined in OSM. Unfortunately, a large amount of these categories have duplicates with different notations or misspellings. The category "jewelry", for example, has seven variations within OSM (e.g., jeweler, jewelers, jewelery, jeweller, jewellers, jewellery, jewelry), which makes a post-filtering of location entities by category extremely difficult. In order to ease the filtering process, a mapping of duplicate entries to a reference category name is performed before importing the data into the LCD dataset. This is done by storing a list of all OSM POI categories into the database and by extending them with a reference category name used for the import process. This reference name is generated for each category by applying *OpenRefine*[7] to its set of possible variations.

In location analytics scenarios, the dwell time for a location visit is a very important piece of context information to utilize since it can give an implicit statement about the popularity of a place. For instance, spending much time in a restaurant might give a positive statement about its popularity, whereas a long stay in a barber shop might be negative. For this purpose, the timestamp of the server is stored in minutes after midnight for the check-in as well as check-out enabling the calculation of dwell times for location visits. In the first version of the app, the user had to manually check-in and out from a location entity. This had the advantage that we were able to collect precise check-in and check-out times if this function was used correctly by the user. However, in the course of time, we have realized that a manual check-out met with little response from a user experience perspective since users tended to forget to check-out from a place in time leading to large and incorrect dwell time calculations.

Therefore, we have implemented an automatic check-out function within our app integrating parts of the *Geofencing SDK* introduced in [126], which is built around *Google's Geofencing API*[8] and ensures precise *entering* as well as *leaving* location events. All location entities have a ground area, which is described by a center position and a radius in meters. With a check-in to a certain place, the center position as well as the radius of this place is used to define a *geofence* (i.e., a polygon that describes a geographic area) around it. The *Geofencing SDK* analyzes the position of the user and determines whether the user is inside or outside of the location entity's geofence. In addition, if available, a complete list of the WiFi APs that are scanned at the moment of the check-in is stored and continuously compared to the current list

---

[7] http://www.openrefine.org/.

[8] http://developer.android.com/training/location/geofencing.html.

of APs measured by the user. As soon as the user leaves the geofence and/or none of the originally scanned APs is seen anymore, an automatic check-out is performed within the app.

There are rare situations in which a smartphone can lose its signal or Internet connectivity for a long period of time while being checked in at a certain location, e.g., at the airport during boarding. In such conditions, the app is not able to perform automatic check-outs, again leading to large dwell times and false data in the database. Here, the server applies filtering mechanisms to reduce faulty values in the LCD dataset. During the import process, the check-in information with large dwell times are hold back. The location category of the POI is queried and compared to a white list that defines which kind of POIs tend to produce large dwell times (e.g., resorts or hotels). If the check-in was performed in such a place, the data is likely to be valid and is therefore stored into the triplestore. Otherwise, the check-out time is automatically reduced to a predefined threshold.

In the near future, we plan to optimize the automatic check-out process when losing signal or Internet connectivity by saving the time and location of the user immediately after the loss happens. As soon as the user is online again, the stored time and location can be compared to his current time and position in order to determine whether a check-out happened in between or not. Furthermore, we will work on utilizing our *Semantic Tracking* approach (see Sect. 7.2.1) for also automatically checking in users when being in frequently visited locations.

Mobile network and WiFi AP topology data is another type of context information that is acquired by the *CDCApp* and is used in a manifold way: Regardless of whether a user is checked in or not, the app constantly runs a background service every three minutes that collects network measurements and sends this information to the *OpenMobileNetwork*. By doing so, we ensure that the context-aware services of the *CDCApp* that mainly rely on the *Semantic Tracking* and the network topology data of the OMN work properly even in geographic areas where the OMN initially had no or less data.

In parallel, for the period of a check-in, the same network measurements are also stored locally on the client every three minutes in combination with the indoor/outdoor value coming from the *IODetector* in order to later relate this check-in to the mobile network cells and WiFi APs that were mainly seen when being at a certain place for a certain amount of time. Additionally, the number of how often a WiFi AP or a mobile network cell has been seen during the check-in is also stored. This is done due to the fact that an automatic check-out is not triggered immediately right after the user has left the place, but rather a few minutes later. In order to avoid the storage of false network topology data that might be collected between the time interval of the user leaving the location and the automatic check-out being triggered, the collected network data is filtered during the check-out process according to the number of how often it was seen. Here, we assume that the set of "correct" mobile network cells and WiFi APs will be seen more often and hence the filtering is done by calculating the ratio of how often a WiFi AP or cell has been seen and by removing all with a ratio lower than a defined threshold. Knowing a fingerprint of WiFi APs and mobile network cells that were measured when being inside a restaurant, for example, enables the mobile network operator to provide specific analytics information for location hot spots to third-party developers and business customers.

The *CDCApp* also collects and processes location preferences in the form of favorite and frequently visited locations. A favorite location is a place that a user actually likes to visit such as his favorite restaurant (see Fig. 5.3). This information is explicitly given by the user by adding a POI to the list of favorites within the app. A frequently visited location, on the other hand, represents a location where he is obligated to spend much time such as his office or school. These locations are determined based on his check-in history and updated with each new check-in. Due to the fact that the check-in behavior of a user is usually different in the week compared to the weekend, the analysis of frequently visited locations is separated into workdays and weekends. At first, the number of check-ins for each POI is counted and divided by the total number of workdays where the user has performed at least one check-in. To avoid scenarios in which a user has checked in at work several times per day due to some breaks in between (e.g., lunch), only one check-in per workday is considered for calculating the ratio. The same procedure also applies for weekends.

For providing human-readable descriptions of the user's frequent check-in behavior, the average workday availability is calculated taking Germany as a reference. An employee has in average 30 vacation days and misses 10 workdays due to illness. Assuming 250 workdays per year, the calculation leads to 210 workdays per user and a ratio of 21/25. This ratio multiplied with the ratio for a visited location that has been calculated beforehand, provides the human-readable information about how often a POI has been visited normally per week. For example, a value above 4/5 would mean that the POI is normally visited more than four times a week leading to a human-readable description of "mostly everyday" within the app. The average weekend availability, on the other hand, is calculated based on the average amount of vacation days of an employee, which yields to 47 regular weekends out of 52 per year without taken holidays.

5.2.3 Context Data Cloud Ontology Design

LCD[9] is based on the comprehensive *Context Data Cloud Ontology* (see footnote 14 in Chap. 1) that consists of various ontology facets (e.g., *Location*, *Context Situation*, or *Additional Information*) representing all aspects of the collected data. This ontology reuses concepts of well-known vocabularies (e.g., *Good Relations Ontology*[10], *DCMI Metadata Terms*[11], *WGS84 Geo Positioning Vocabulary*, or *vCard Ontology*[12]) and introduces new concepts with `cdc-owl` as a prefix.

---

[9]http://www.contextdatacloud.org/lcd/.

[10]*gr*, http://purl.org/goodrelations/v1#.

[11]*dcterms*, http://dublincore.org/documents/dcmi-terms/.

[12]*vcard*, http://www.w3.org/TR/vcard-rdf/.

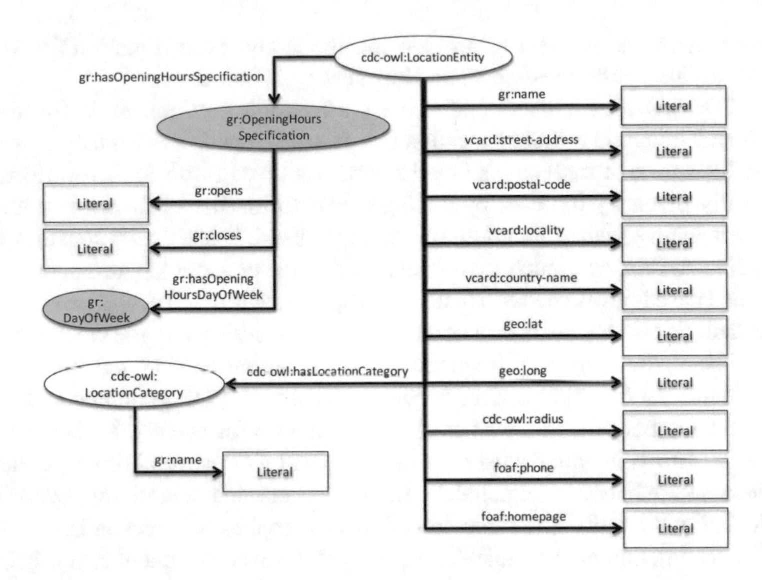

Fig. 5.4 Context Data Cloud Ontology – location facet

Figure 5.4 shows the *Location* facet of the *Context Data Cloud Ontology*. This facet is similar to the data provided by *LinkedGeoData*, for example, and comprises static information attached to a `cdc-owl:LocationEntity`, which describes a generic concept for an object related to a location (e.g., a POI or an event).

A `cdc-owl:LocationEntity` has its name as a textual representation (`gr:name`), its address (`vcard:street-address`, `vcard:postal-code`, `vcard:locality`, `vcard:country-name`), the geo coordinates of its position (`geo:lat`, `geo:long`), its ground area defined by a radius in meters (`cdc-owl:radius`), its phone number (`foaf:phone`), and homepage (`foaf:homepage`) attached to it. Opening hours for this `cdc-owl:Location Entity` are set by the `gr:OpeningHoursSpecification` concept, whereas the property `cdc-owl:hasLocationCategory` defines a category for it (e.g., restaurant or airport).

The *Location* ontology facet is closely connected to the *Context Situation* facet (see Fig. 5.5) that relates a user's `cdc-owl:CheckIn` including the collected context parameters to a `cdc-owl:LocationEntity` via the `cdc-owl:refersTo` property.

The dwell time for a location visit is given by a difference calculation between the minutes after midnight values of the `cdc-owl:hasCheckInMinutes` and `cdc-owl:hasCheckOutMinutes` properties, whereas the day of the `cdc-owl:CheckIn` is represented by its `dcterms:date` and a textual representation of the `gr:DayOfWeek`. Contextual information is automatically collected during the visit (e.g., weather condition, holiday information, or network data) or via manual input (e.g., rating or comment) and is used to enrich the

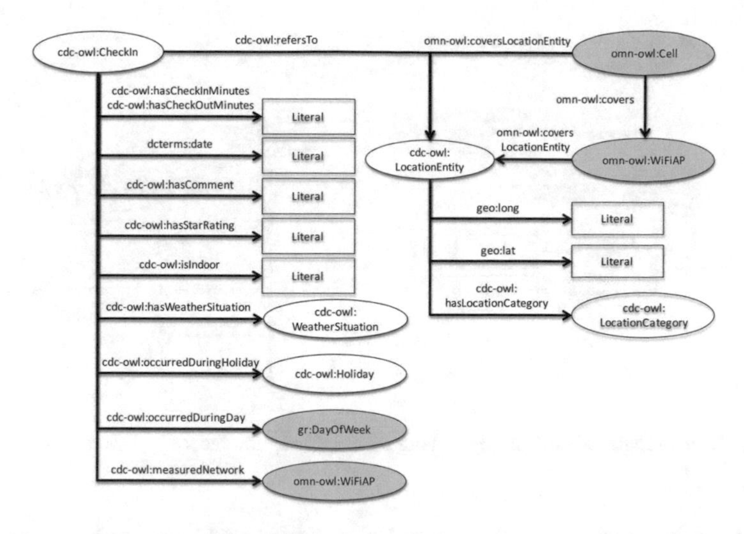

Fig. 5.5 Context Data Cloud Ontology – context situation facet

`cdc-owl:CheckIn` data. Users are able to numerically rate (`cdc-owl:hasStarRating`) and comment (`cdc-owl:hasComment`) the check-in, for example. The boolean value for the `cdc-owl:isIndoor` property denotes whether the user has spent his time in an indoor or outdoor environment during the check-in time, while the relation to the `cdc-owl:WeatherSituation` concept describes the current weather condition at the moment of the check-in. If the user checked in to a certain place on a public holiday, an instance of the `cdc-owl:Holiday` concept is created and connected to the check-in using the `cdc-owl:occurredDuringHoliday` property. In addition, all WiFi access points and mobile network cells including the neighboring cells that the smartphone was able to measure during the check-in process are attached to the check-in (`cdc-owl:measuredNetwork`) through a direct interlink to the OMN. In contrast to the property `cdc-owl:measuredNetwork`, the `omn-owl:coversLocationEntity` predicate is used on the OMN side to geographically relate the coverage area of a mobile network cell or WiFi AP to the `cdc-owl:LocationEntity`. Via this link, all location entities that have a certain number of check-ins for a specific historic context situation, for example, can be resolved through the OMN as soon as the user enters or is in the vicinity of the coverage areas of these cells or APs (see Sect. 8.2.2).

The list of contextual information is not considered complete and can be extended as required. In the near future, we plan to include *tags* (e.g., business lunch, dinner, or birthday) that enable a more fine-grained and user-generated categorization of the location entities. We also consider to collect more sensor data directly from the smartphone.

All kinds of location-related crowdsourced data that is not considered as the current context of a check-in, is modeled in the *Additional Information* ontology facet. This could incorporate the dishes offered by a restaurant that are rated by

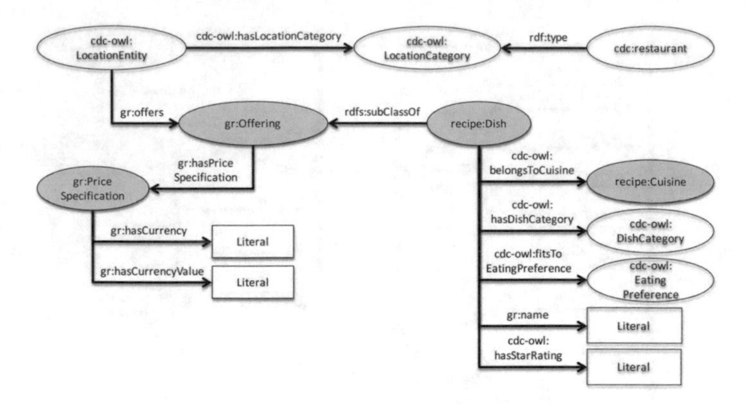

Fig. 5.6 Context Data Cloud Ontology – additional information facet

users in terms of taste or special vegetarian groceries sold in a certain grocery store, for example. Having this kind of additional information enables more sophisticated location analytics scenarios in which not only aspects of the coarse location itself are taken into consideration, but also detailed information such as concrete offerings. By using such kind of data, the network operator becomes capable of realizing use cases such as the "most popular restaurants *offering pasta below 10 euros* where people tend to go on weekends in the afternoon".

Figure 5.6 highlights the *Additional Information* facet that includes exemplary concepts for representing restaurant dishes offered by a `cdc-owl:Location Entity`, which has `cdc:restaurant` as a `cdc-owl:LocationCategory`. Currently, this facet mainly incorporates concepts of the *Good Relations Ontology* for describing `gr:Offerings` as well as `gr:PriceSpecifications` and concepts of the *Linked Recipes*[13] vocabulary for representing dishes.

A `cdc-owl:LocationEntity` `gr:offers` a `gr:Offering`, which is the superclass for all types of offerings. `recipe:Dish` is an `rdfs:subClassOf` `gr:Offering` and describes a certain dish provided in a restaurant. This dish has a textual representation of its name (`gr:name`) and belongs to a certain `recipe:Cuisine` as well as a `cdc-owl:DishCategory` (e.g., pasta). The `cdc-owl:fitsToEatingPreference` property defines the suitability of this dish for people with special `cdc-owl:EatingPreferences` (e.g., vegetarian), whereas the `cdc-owl:hasStarRating` predicate relates a numerical rating value to it. A certain price (`cdc-owl:hasCurrencyValue`) can be added to the dish by using the `gr:PriceSpecification` concept.

In order to enable location analytics scenarios that make use of such location-related domains (e.g., groceries in grocery stores or movies in movie theaters), we assume that within this facet, new concepts providing additional information to those domains are modeled by the respective domain expert.

[13] *recipe*, http://linkedrecipes.org/schema.

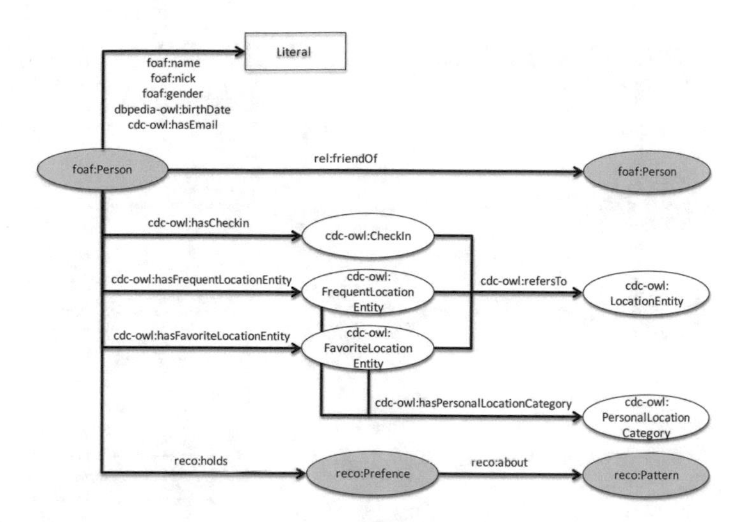

Fig. 5.7 Context Data Cloud Ontology – user profile facet

Data that is populated using the *Location*, *Context Situation* and *Additional Information* facets, is publicly available through our SPARQL endpoint. Due to privacy reasons, this data cannot be traced back to certain users. The endpoint only provides a set of check-ins made within a certain context without relating this information to users.

However, the *User Profile* ontology facet (see Fig. 5.7) that is hosted on a private SPARQL endpoint[14] creates this link between user (`foaf:Person`) and `cdc-owl:CheckIn` via the `cdc-owl:hasCheckIn` property and allows the platform providers to personalize their location analytics services.

A `foaf:Person` is represented by basic user information properties such as his name (`foaf:name`), his nickname (`foaf:nick`), his gender (`foaf:gender`), or birthday (`dbpedia-owl:birthDate`). Social relationships to other users are described using concepts from the *RELATIONSHIP* vocabulary. One example for a relationship is the `rel:friendOf` predicate.

Furthermore, domain-specific preferences are defined based on the concepts of the *REcommendations Ontology*. According to [117], a `foaf:Person` `reco:holds` a `reco:Preference` (e.g., eating preference) with a certain `reco:Pattern` (e.g., being vegetarian and/or having a price limit). These preferences and their corresponding ontology concepts are usually imported into the user profile with every registered service on the platform. For example, if the user activates the *Restaurant Recommender* within the *Popular Places Finder* service of the *CDCApp*, concepts for describing eating preferences are imported automatically into the user profile enabling the user to provide the required information.

[14]http://contextdatacloud.org:8891/sparql-auth.

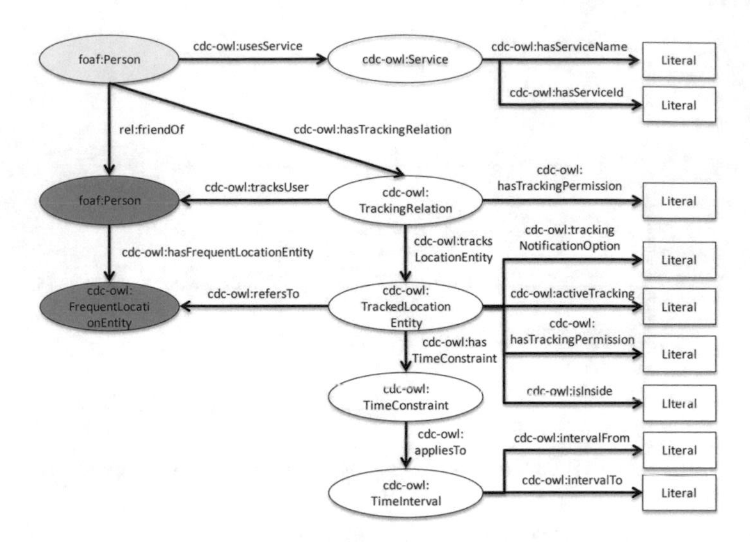

Fig. 5.8 Context Data Cloud Ontology – tracking and service facets

In addition to domain-specific interests, location preferences in the form of `cdc-owl:FavoriteLocationEntitys` or `cdc-owl:Frequent LocationEntitys` are also added to the user profile. Both types of location preferences can be personally categorized (e.g., *T-Labs* is the user's *office*) using the `cdc-owl:has PersonalLocationCategory` property. This information is crucial in location analytics scenarios that analyze the popularity of places based on a huge number of people favoriting locations, for example.

To complete the picture, Fig. 5.8 shows the *Tracking and Service* ontology facets that are mainly used for our *Semantic Tracking* approach presented in Sect. 7.2.1 with the goal of setting privacy-compliant location tracking constraints between two tracking parties.

A `foaf:Person` (light brown), who is the friend of another `foaf:Person` (dark grey), uses a tracking service (`cdc-owl:Service`) with a name (`cdc-owl:hasServiceName`, e.g., Friend Tracker) and a service ID (`cdc-owl:hasServiceId`) running on his device. The social relationship (`rel:friendOf`) between both users is set via a basic friend/family status request and a confirmation allowing the `foaf:Person` (light brown) to send a tracking request in the next step. With an accepted tracking request, a one-way `cdc-owl:TrackingRelation` between those two friends is created, which enables a general tracking of the `foaf:Person` (dark grey). However, the `cdc-owl:TrackingRelation` includes a permission flag (`cdc-owl:has TrackingPermission`) that can be set by the `foaf:Person` (dark grey) in order to permit or forbid being tracked. Is tracking permitted, the `foaf:Person` (light brown) is capable of seeing the favorite and frequently visited location entities of his friend and is able to select which ones he would like to track. For each location entity to be tracked, a `cdc-owl:TrackedLocationEntity` resource is built with a reference to the `cdc-owl:FavoriteLocationEntity` or `cdc-owl:FrequentLocationEntity` and other parameters for the tracking party: `cdc-owl:trackingNotificationOption` enables the `foaf:Person` (light brown) to adjust the notification options (e.g., entering, leaving, or on-change), whereas `cdc-owl:activeTracking` allows him to activate or deactivate tracking. In addition, the `foaf:Person` (dark grey) can reduce the tracking to certain locations (`cdc-owl:hasTrackingPermission`) or to a time interval (`cdc-owl:TimeConstraint`).

5.3 OpenMobileNetwork Geocoding Dataset

As described in Sect. 7.2.2, geocoding is the process of deriving WGS84 coordinates for a location specified by a user in a human-readable form. For our exemplary *Semantic Geocoding* service, we created the *OpenMobileNetwork Geocoding Dataset* containing address data and interlinked it with the OMN as another context source for optimizing geocoding results (through the usage of network topology data) even if the address query itself is less precise. The dataset is made available through a SPARQL endpoint[15] and comprises address data for Berlin, San Francisco, and Helsinki with 397,269 buildings and 22,184 streets in total.

In order to enrich a geocoding query with a user's coarse location context (being the mobile network cell he is connected to), it is required to geographically map address data to mobile network topologies. For this purpose, we utilized the geographic mapping model of the OMN (as introduced in Sect. 5.1.1) and extended it with address data as illustrated in Fig. 5.9.

For applying the geographic mapping model, address data needs to be structured to its unique entities, provided in *Linked Data* format and described in an ontology for enabling an interlinking with the OMN. LGD already provides structured address data in RDF. However, in order to ensure up-to-dateness of data and permanent availability, we decided to extract address data directly from OSM instead of LGD. Two data types of OSM are mainly relevant for our purposes. The first type is a `node`, which stands for a geographic point. Besides being a geographic position, a node can also be a house, for example. A number of nodes can form a `way` that represents a line or a polygon building structures, regions, or streets. Address information in nodes and ways are incorporated using `tag` elements, e.g., `addr:street` and its corresponding value.

Based on the OSM data model, we defined the *OMN Geocoding Ontology*[16] as an extension of the *OMN Ontology*.

[15] http://www.openmobilenetwork.org/geocoding/sparql.

[16] *omng-owl*, http://www.openmobilenetwork.org/geocoding/ontology/.

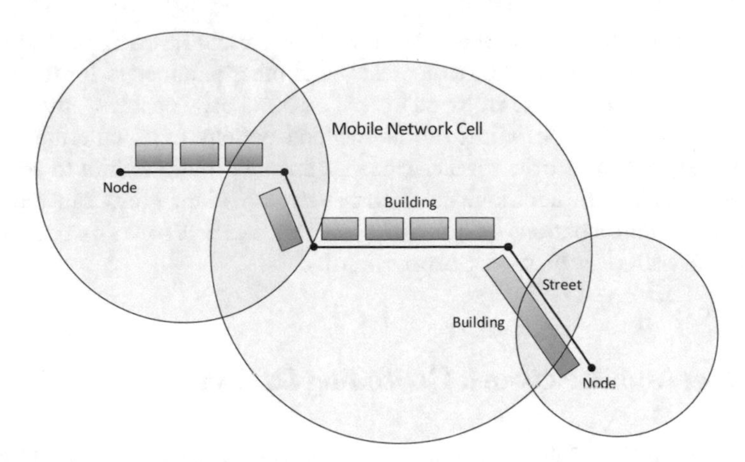

Fig. 5.9 Geographic mapping of address data to network topology data

5.3.1 OMN Geocoding Ontology

The *OMN Geocoding Ontology* (see Fig. 5.10) uses `omng-owl` as its prefix for ontology concepts and `omng` for resources. It further reuses concepts of the *WGS84 Geo Positioning Vocabulary* and the *ISA Programme Location Core Vocabulary*.[17]

Three main concepts are included within the ontology that are based on the OSM data model. The concept `omng-owl:Street` incorporates the predicate `locn:thoroughfare` representing the street name, `locn:postCode` standing for the postal code, and the `locn:adminUnitL1` as well as `locn:adminUnitL2` properties describing the country and the state. Due to the fact that for the mobile geocoding use case (as introduced in Sect. 7.2.2), the search is narrowed to nearby located streets or house numbers, there is no need to index administrative area names as own entities. Instead, this information is stored in the `locn:postCode`, `locn:adminUnitL2`, and `locn:adminUnitL1` properties. An `omng-owl:Street` `omng-owl:consistsOf` a number of `omng-owl:Nodes`, which are geographic points with latitude (`geo:lat`) and longitude (`geo:long`) values. `omng-owl:Building`, on the other hand, is a concept representing a house or a building located on a street. It has a house number (`locn:locatorDesignator`) as well as a geo position and `omng-owl:belongsTo` to a certain `omng-owl:Street`.

An `omng-owl:Node` and an `omng-owl:Building` are linked to mobile network and WiFi AP topology data in the OMN via the `omng-owl:isCoveredByWiFiAP`, `omng-owl:isCoveredByCell`, and `omng-owl:isCoveredByLAC` properties. This enables the search for house numbers or parts of streets covered by certain WiFi access points, mobile network cells, or LACs that are shared among mobile network cells in a specific area.

[17] *locn*, http://www.w3.org/ns/locn.

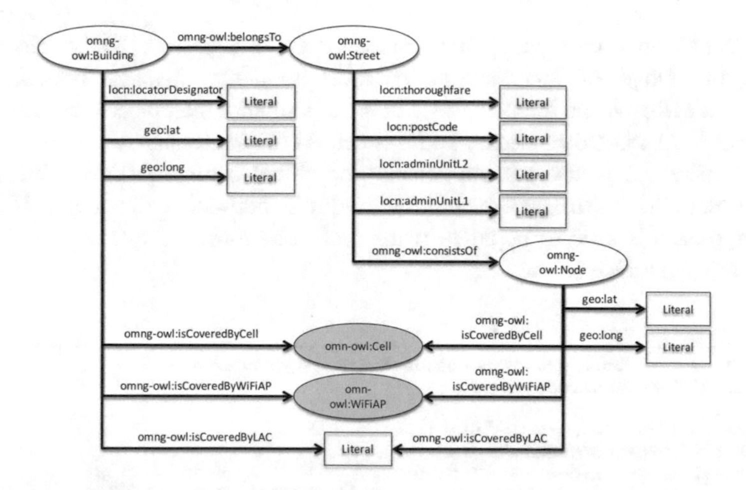

Fig. 5.10 OpenMobileNetwork Geocoding Ontology

5.3.2 Address Data Extraction and Semantification

For three exemplary cities (e.g., Berlin, Helsinki, and San Francisco), we extracted all nodes and ways from OSM including the relevant tag elements. We ignored duplicates (e.g., a node for a store that is also modeled as a house) and nodes with completely missing address information.

In order to populate our ontology model, we needed to have a set of full addresses. However, due to the fact that OSM data is based on crowdsourcing, address information is often scattered across various entities or is partially missing. For this purpose, we used the concept of *guessing tables*, which support the extraction of more complete addresses from OSM data by guessing missing tag fields at locations with similar information. For example, consider two houses with almost exact address information except for a different house number and a missing city name for the second house. Since both houses have the same street name, the same postal code, and the same country name, we could guess that both houses are located in the same city and fill in the missing tag field. A guessing table is populated with postal codes, city names, and other higher level administrative area names for street segments where this information is provided. If another street segment with the same name is tagged differently, guessing is disabled for this street name, i.e., there is no possibility to derive information from it.

All street segments with the same administrative area names and postal codes are merged into resources of the `omng-owl:Street` concept. `omng-owl:Nodes` are populated with all points of the joint street segments, whereas values for `omng-owl:Buildings` are split of from the street segments. For each house number, a resource of `omng-owl:Building` is created and the `omng-owl:belongsTo` predicate is set to point to the corresponding `omng-owl:Street`.

Whenever an omng-owl:Building or an omng-owl:Node resource is within the range of an omn-owl:Cell (i.e., the distance between the omng-owl:Node and the omn-owl:Cell is smaller than its coverage area), the link omn-owl:isCoveredByCell is set. At the same time, an omn-owl:isCoveredByLAC is also created causing individual omng-owl:Buildings or omng-owl:Nodes to be covered by multiple network cells and LACs. The same approach is used to populate omn-owl:isCoveredByWiFiAP links for omn-owl:WiFiAP entities.

```
PREFIX omn: <http://www.openmobilenetwork.org/resource/>
SELECT DISTINCT ?street ?name
WHERE {
  ?street rdf:type omng-owl:Street .
  ?street locn:thoroughfare ?name .
  ?street omng-owl:consistsOf ?nodes .
  ?nodes omng-owl:isCoveredByCell omn:cell2160591_5275_262_1
}
```

Listing 5.4 Querying *www.openmobilenetwork.org/geocoding/sparql* for street names that are in the coverage area of omn:cell2160591_5275_262_1

Listing 5.4 illustrates the interlinking between OMN and the address data through an exemplary SPARQL query. By executing this query, all street names are listed that are in the coverage area of the omn:cell2160591_5275_262_1.

Chapter 6
OpenMobileNetwork – A Platform for Providing Semantically Enriched Network Data

This chapter gives an insight into the implementation of the *OpenMobileNetwork* providing an end-to-end view of all platform components. Section 6.1 highlights the system architecture as a direct result of the functional architecture, whereas Sect. 6.2 focuses on the smartphone clients that were developed for network context data crowdsourcing. In Sect. 6.3, on the other hand, the backend functions and the incorporated processes are discussed and demonstrated in detail.

Please note that *Linked Crowdsourced Data* (see Sect. 5.2) as well as the associated *CDCApp* client including the *Semantic Tracking* service (see Sect. 7.2.1) are deployed on the *Context Data Cloud (CDC)*[1] platform that utilizes the *OpenMobileNetwork* in a manifold way. However, the implementation of LCD as well as the *OpenMobileNetwork Geocoding Dataset* (see Sect. 5.3) is quite equivalent to the realization of the *OpenMobileNetwork*. Furthermore, the functional entities as well as the system components of the CDC platform are not in the scope of this work. Therefore, we do not present these aspects within this thesis.

6.1 System Architecture

Figure 6.1 illustrates the system architecture of the *OpenMobileNetwork* as a result of mapping the functional entities described in Sect. 4.2.2 to system components.

In order to enable the crowdsourcing of network context data, we have implemented several *Android* smartphone apps serving different purposes that collect relevant data in the form of network measurements. These network measurements are sent to the *OpenMobileNetwork* backend, which has been realized

[1] http://www.contextdatacloud.org/.

A. Uzun, *Semantic Modeling and Enrichment of Mobile and WiFi Network Data*, T-Labs Series in Telecommunication Services,
https://doi.org/10.1007/978-3-319-90769-7_6

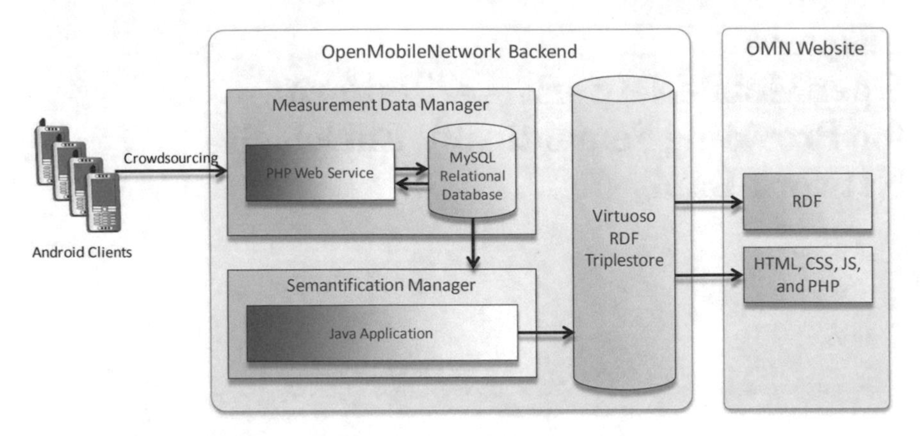

Fig. 6.1 OpenMobileNetwork – system architecture

using different technologies for network context data collection, processing, and semantification.

The *Measurement Data Manager* that is responsible for acquiring the incoming raw measurements and estimating positions as well as coverage areas, is implemented using *PHP 5.3.1*[2] (due to its simplicity in programming) and *MySQL 5.5.38*[3] as a relational database. The *Semantification Manager*, on the other hand, that "semantifies" the estimated network topology data according to the principles of *Linked Data* and interlinks it to other datasets in the *LOD Cloud*, is developed in *Java 1.6*[4] in combination with the *Apache Jena 2.6.4*[5] framework and uses a *Virtuoso 7.1* triplestore for storing and publishing the semantically enriched and interlinked network topology data. The connection to the triplestore is established by using the *Virtuoso Jena Provider*[6] in the Java application.

Two separate data storages are comprised within the architecture design. The MySQL relational database as part of the *Measurement Data Manager* is responsible for storing the crowdsourced raw measurements as well as the approximated network topology data after the position and coverage area estimation process, whereas the semantically enriched data is transferred into a Virtuoso triplestore (with a little delay) after the semantification and interlinking process (being functions of the *Semantification Manager*).

By doing so, we ensure that the raw measurements, which can cause a potential privacy issue when theoretically exploited for inferring user movements, are not freely accessible via an endpoint. Secondly, we separate the data access for the calculation operations with potentially competing access to data coming from end

[2]http://www.php.net/.

[3]https://www.mysql.com/.

[4]https://www.java.com/.

[5]https://jena.apache.org/.

[6]https://github.com/srdc/virt-jena.

users. This decreases the total load of the database servers. Another reason for using a relational database in parallel to a triplestore is that over the years, relational databases reached a very high performance and maturity level. In addition, they are well-known by developers and easy to use in the implementation process of a proof of concept platform. Therefore, we decided to handle the storage and processing of the raw measurements in such a database (also serving a backup purpose). For publishing the data, we pragmatically applied the principles of *Linked Data* and hence utilized semantic technologies.

Classic Web technologies, such as HTML,[7] CSS,[8] JavaScript (JS),[9] and PHP, are further utilized for implementing the *OpenMobileNetwork* frontend website (including the HTML and RDF descriptions of the data resources) as well as the visual coverage map that shows mobile network cells, WiFi access points, their coverage areas, and further detailed information when clicking on them.

6.2 Smartphone Clients for Network Context Data Collection

Section 4.3 highlighted several crowdsourcing methods for collecting network context data. In this section, we introduce the implemented apps that apply these methods. Section 6.2.1 presents the *OpenMobileNetwork for Android* app, which uses a systematic wardriving and warwalking method, whereas Sect. 6.2.2 describes the location-based game *Jewel Chaser* that applies a gamification approach for collecting network context data. The *Context Data Cloud for Android* app, on the other hand, is introduced in Sect. 6.2.3 as an example for a crowdsourcing background service that runs in an app with another purpose.

6.2.1 OpenMobileNetwork for Android (OMNApp)

OpenMobileNetwork for Android (see footnote 8 in Chap. 4) (see Fig. 6.2) is a network context data collection app in the version 0.7 running on *Android 2.2* and higher, which is expected to be used by performing systematic wardriving and warwalking. It actively collects network context data in order to estimate and derive mobile network cell and WiFi AP positions. In addition, the app also acquires dynamic network context information, such as the total traffic produced on smartphones or service usage information, in order to enable the modeling and visualization of the current and historic state of the network.

---

[7]https://www.w3.org/html/.

[8]https://www.w3.org/Style/CSS/.

[9]https://www.javascript.com/.

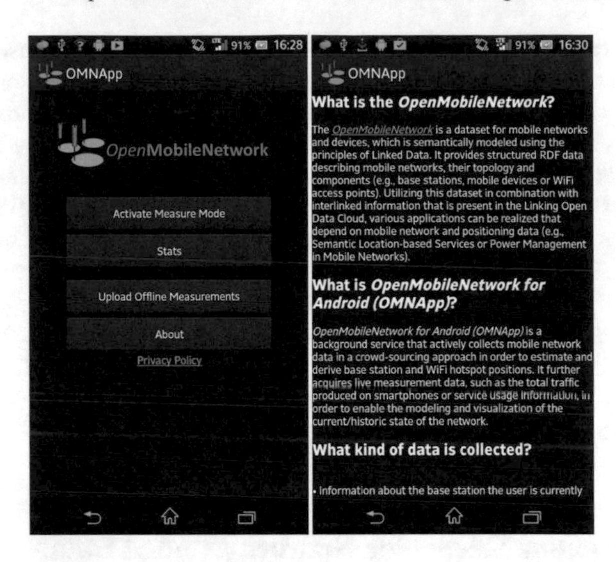

Fig. 6.2 OpenMobileNetwork for Android (OMNApp)

The app provides two operation modes. Based on the availability of capturing nearby WiFi access points on the device, the measurement modes are automatically selected.

In the *WiFi Measure Mode*, the GPS module is immediately activated and listens for location changes. Moreover, a service is started every 15 s, which forces the WiFi module to scan for access points in its reach. After receiving the results of the scan, the last location data along with the mobile network topology and dynamic network context information is stored and gathered as one measurement. A tolerance of three seconds between WiFi scans and the GPS location fix has been chosen to keep the drift between both data at a minimum level. The second mode is selected if WiFi is not available on the device. Every 30 s, the *Active Measure Mode* service is started in which the measurement of the mobile network is fully dependent on the GPS module. As soon as the GPS location fix is ready, mobile network topology in combination with dynamic network context data is captured and stored.

After having a complete measurement, *OMNApp* checks if the device is online and decides on whether the measurement is written on the internal/external storage or if it can be transferred to the *OpenMobileNetwork* server directly. This procedure ensures that no measurement is lost even if there is no Internet connection and provides the option to contribute to the *OpenMobileNetwork* without causing cellular traffic. Measurements that have been stored locally can be uploaded later on at anytime when Internet is available again (e.g., when connected to the WiFi access point at home) by clicking on the *Upload Offline Measurements* button.

OMNApp runs in the background once activated meaning that users can use their smartphone as usual without being disturbed by the app. This enables the acquisition of authentic dynamic network context data such as information about service usage or incoming and outgoing traffic. However, due to the fact that this app is designed to be used via wardriving and warwalking without providing a real incentive for the common app user, it is highly difficult to motivate a wide range of users to contribute with their data.

6.2.2 *Jewel Chaser*

As a proof of concept for crowdsourcing via gamification, the location-based game *Jewel Chaser* (see footnote 10 in Chap. 4) has been developed that runs on *Android 2.2* or higher and uses the *OMN Measurement Framework* introduced in Sect. 4.3.2.1 in its core. Figure 6.3 shows the map and jewel collection screens as two screenshots of the app.

In this game, the goal of the player is to collect different types of jewels that are virtually situated in locations of the real world. In order to play the game, the player needs to click on the *Start Collecting* button on the start screen, check the map, and move towards one of the jewels displayed on it. As soon as the player reaches a location where a jewel is placed, the type of the jewel is revealed in the form of a picture and added to the collection of the player. Moreover, the player is rewarded with points and he is ranked on a high score based on the number of jewels he has collected.

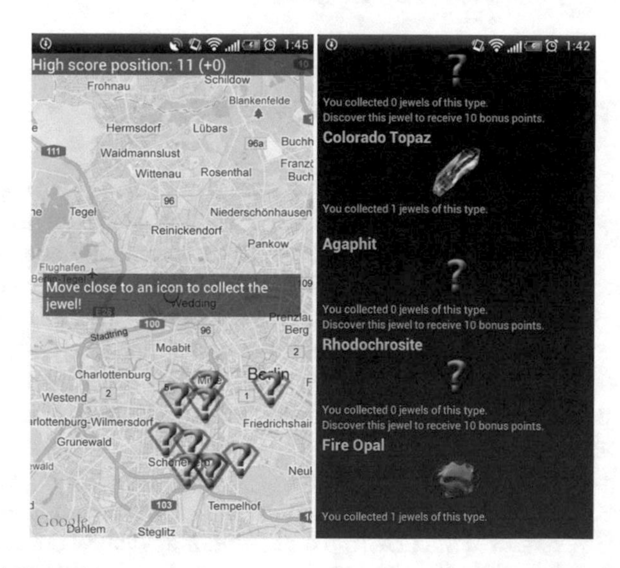

Fig. 6.3 Jewel Chaser

Jewel Chaser utilizes both functions of the *OMN Measurement Framework*. In order to motivate the user to move to points where measurements are needed, *Jewel Chaser* uses the framework to obtain FMLs from the OMN as soon as the user starts the game. Having a set of FMLs out of which a list of geo points are selected by the framework, the game places the jewels according to the selected geo points (which most likely correlate to the center of the provided FMLs). The user can view the locations of the jewels on a map within the game and move towards the closest one. As soon as the player reaches one of the points where a jewel was placed, a full network context measurement is conducted including mobile network as well as WiFi access point data.

In order to gather the maximum amount of data, *Jewel Chaser* does not only conduct measurements once the player reaches the target location, but continuously collects data while the player has the game open. Measurements are acquired every time the player moves at least 100 m. This allows to continuously gather measurements while the player is on his way to a jewel, but ensures that the server is not flooded with unchanged data. Due to the game design decisions discussed in Sect. 4.3.2.2, data is only acquired when the user keeps the game running and is on the game map.

Jewel Chaser is designed to collect measurements in a distributed fashion and users are expected to play the game actively with GPS functionality enabled. However, the game is not very suitable for applying wardriving methods since it is designed for pedestrians and jewels are sometimes placed in locations a car cannot reach. Moreover, the fact that data is not conducted as soon as the app is closed might decrease the number of measurements compared to a service collecting data in the background, which the user does not need to explicitly run in the foreground.

6.2.3 Context Data Cloud for Android (CDCApp)

The *Context Data Cloud for Android* (see footnote 11 in Chap. 4) [C6] app is available at *Google Play* since August 2014 and the latest 1.4.3 version runs on smartphones with *Android 4.0* and higher. It is a location-based community app with the purpose of collecting place preferences of users as well as additional location-related information in combination with their contextual situation. This app offers a set of semantically enriched services to users, such as a *Friend Tracker* [C10] or a *Popular Places Finder* (see Sect. 8.2), and further enables users to check-in to certain POIs in their vicinity providing information about their location visits to other users within their social community.

By comprising a crowdsourcing background service for collecting network measurements, the *CDCApp* is also part of our holistic crowdsourcing strategy. In the first place, we implemented the *CDCApp* for contributing diverse context data to the LCD dataset in order to interlink this information with the network topology data of the OMN (see Sect. 5.2 for further details). However, as a positive side effect, mobile network and WiFi AP topology data is also acquired by the *CDCApp* and is used in a manifold way. For this purpose, the app constantly runs a background service every three minutes that collects network measurements (including the current list of scanned WiFi APs, the mobile network cell the user is connected to as well as the measured neighboring cells) and sends this information to the OMN for indirectly enriching its dataset of approximated network topology data. By doing so, we ensure that the context-aware services of the *CDCApp* that mainly rely on the *Semantic Tracking* (see Sect. 7.2.1) and the network topology data of the OMN work properly even in geographic areas where the OMN initially had no or less data. In addition, we enrich the OMN dataset with new measurements.

In contrast to the other apps mentioned in Sects. 6.2.1 and 6.2.2, the *CDCApp* is not designed to always keep GPS enabled. A lot of the app functions, such as the dwell time calculation during a stay at a place, the automatic check-out, or the network measurement collection, run constantly in the background. Keeping GPS always enabled will drain the smartphone battery to a great extent. Furthermore, users do not always reside in outdoor environments (when using the app), so that GPS will not work all the time. Therefore, in order find a good balance between high accuracy estimations and more coverage in the *OpenMobileNetwork*, we also utilize the *Network Location Provider* of *Android* for collecting network measurements if GPS is disabled. However, due to the fact that this leads to measurements with a lower accuracy compared to the ones conducted by the other apps, we flag the cells (and WiFi APs) in the triplestore that are estimated by using these (less accurate) measurements with an `omn-owl:providedByCDC` predicate and replace the estimations as soon as *OMNApp* or *Jewel Chaser* deliver more accurate measurements for the same cell.

As soon as the app is installed on the smartphone, the background service of the app starts collecting data regardless of whether the app is running in the foreground or not. This leads to a big number of network measurements. However, such a crowdsourcing approach comes with several drawbacks: One disadvantage is that data is constantly collected every three minutes regardless of whether the user is moving or not. This means that a lot of redundant data is collected while the user is sleeping, for example. Here, filtering mechanisms (e.g., filter measurements that are redundant) or distance thresholds (e.g., only collect data when the user is moving a certain distance) could increase the quality of the data. Another problem is that the battery consumption of the phone increases if network connectivity is used so often, which will potentially lead to the fact that the user will uninstall the app. In a professional app development environment, these drawbacks need to be taken into consideration. However, due to the fact that these drawbacks are not in the scope of this thesis, we did not tackle them.

Please note that all app details concerning LCD are illustrated in Sects. 5.2.1 and 5.2.2.

6.3 Backend Server

As described in Sect. 6.1, the *OpenMobileNetwork* backend consists of two main functions that are based on different technologies.

The *Measurement Data Manager* is based on a PHP Web service for managing incoming network measurements and keeping the dataset up-to-date. Measurements received from the smartphone clients, such as *OMNApp* (see Sect. 6.2.1), are immediately checked for valid values. The raw data is then written into the MySQL relational database (RDB).

In a second step, a script, which is performed as a cronjob, selects the (newly added) rows within the RDB and approximates the (mobile and WiFi) network topologies out of the raw data based on the selected algorithms (see Sect. 4.4).

After calculating the positions and coverage areas as well as the neighbor relations and service usage information of the mobile network cells and WiFi APs, the *Semantification Manager* function implemented in Java is triggered in order to transform the relational data into RDF according to the principles of *Linked Data*. During this phase, the collected network traffic and user data (see Sect. 4.5.1.2) is also transferred into the needed format. The resulting RDF data is stored in the Virtuoso triplestore and is made available via a SPARQL endpoint.

Each module within both functions can be executed independently according to the actual needs and recognizes if an update of an existing dataset is needed. This leads to maximum flexibility and improves the overall performance. Furthermore, the implementation is integrated into a time-based scheduler for automatic and regular updates of the network topology information within the relational database and the Virtuoso triplestore. This ensures up-to-date data and reduces the effort of manual maintenance of the dataset.

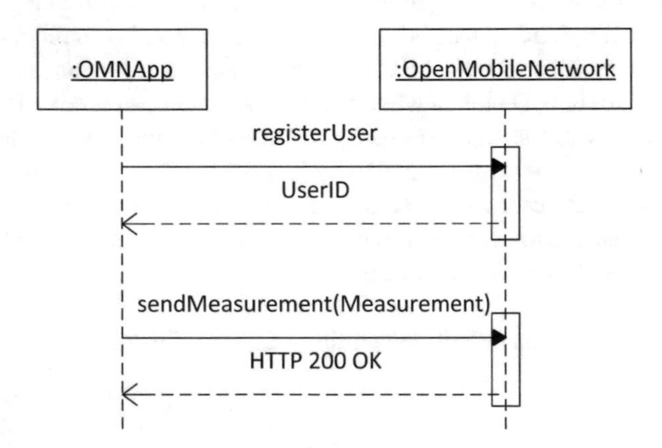

Fig. 6.4 OMN Measurement Data Manager – client-server communication

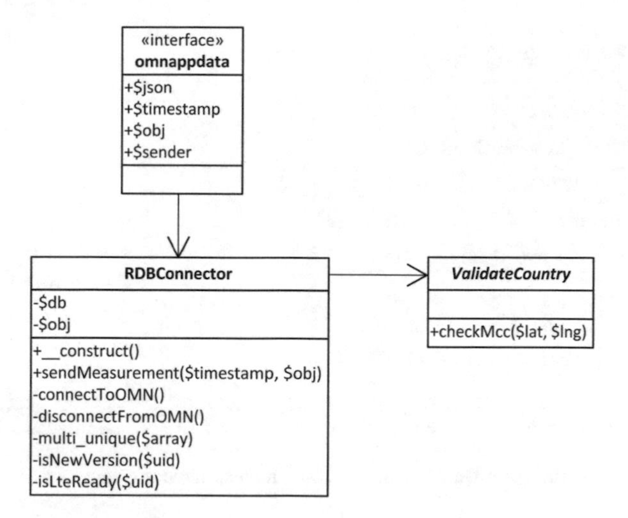

Fig. 6.5 OMN Measurement Data Manager – client connector

6.3.1 Measurement Data Manager

As illustrated in Fig. 6.4, the communication between the respective smartphone clients (here exemplary referred to *OMNApp*) and the *OpenMobileNetwork* server is based on `HTTP POST` requests.

When using a smartphone client for the first time, it will register the user by receiving a user ID from the server. This process is also performed whenever the app gets an update or the user changes the network operator. The user ID is stored within the *SharedPreferences*[10] of the *Android* phone.

As soon as a measurement is ready for transmission (either in online or offline mode), it is combined with the user ID and sent to the server as a JSON[11] object via `HTTP POST`. The server instantly checks the data and, if valid, inserts it into the relational database. An `HTTP 200` status code is returned when storing the data was successful. With this information, the app can safely delete the transmitted data.

Due to the fact that the smartphone clients were implemented in different periods of time and were integrated into the live system at a later stage, the interface(s) for sending those measurements vary slightly and are designed for each smartphone app specifically. Each interface is located at http://www.openmobilenetwork.org/measurement/app_name/app_namedata.php, where `$app_name$` stands for the name of the app. JSON objects of the *OMNApp* measurement data, for example, are sent to the interface http://www.openmobilenetwork.org/measurement/omnapp/omnappdata.php.

---

[10] https://developer.android.com/reference/android/content/SharedPreferences.html.

[11] http://www.json.org/.

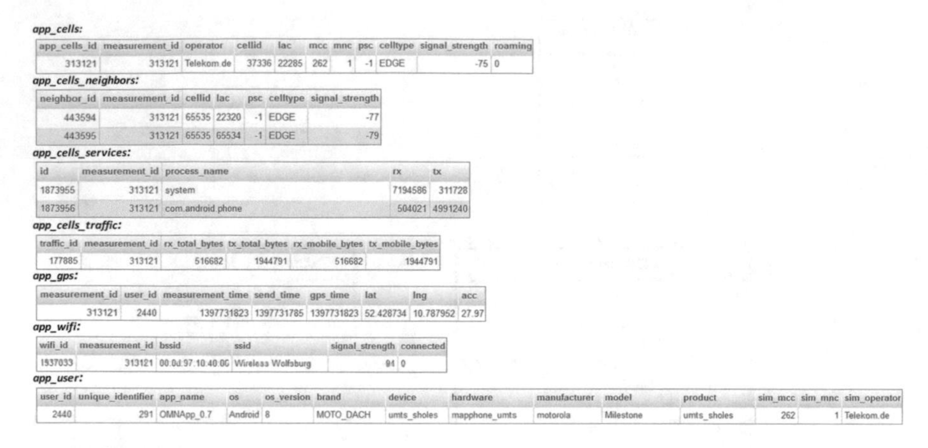

app_cells:

app_cells_id	measurement_id	operator	cellid	lac	mcc	mnc	psc	celltype	signal_strength	roaming
313121	313121	Telekom.de	37336	22285	262	1	-1	EDGE	-75	0

app_cells_neighbors:

neighbor_id	measurement_id	cellid	lac	psc	celltype	signal_strength
443594	313121	65535	22320	-1	EDGE	-77
443595	313121	65535	65534	-1	EDGE	-79

app_cells_services:

id	measurement_id	process_name	rx	tx
1873955	313121	system	7194586	311728
1873956	313121	com.android.phone	504021	4991240

app_cells_traffic:

traffic_id	measurement_id	rx_total_bytes	tx_total_bytes	rx_mobile_bytes	tx_mobile_bytes
177885	313121	516682	1944791	516682	1944791

app_gps:

measurement_id	user_id	measurement_time	send_time	gps_time	lat	lng	acc
313121	2440	1397731823	1397731785	1397731823	52.428734	10.787952	27.97

app_wifi:

wifi_id	measurement_id	bssid	ssid	signal_strength	connected
1937033	313121	00:0d:97:10:40:06	Wireless Wolfsburg	94	0

app_user:

user_id	unique_identifier	app_name	os	os_version	brand	device	hardware	manufacturer	model	product	sim_mcc	sim_mnc	sim_operator
2440	291	OMNApp_0.7	Android	8	MOTO_DACH	umts_sholes	mapphone_umts	motorola	Milestone	umts_sholes	262	1	Telekom.de

Fig. 6.6 OMN Measurement Data Manager – RDB measurement example

Figure 6.5 illustrates the class diagram for the *Client Connector* of *OMNApp* (named as `RDBConnector` within the implementation). `omnappdata` creates an instance of `RDBConnector`, which is responsible for establishing the connection to the RDB. For an incoming network measurement, `RDBConnector` checks the MCC, the MNC, the Cell-ID, the LAC, and the user ID for valid values (e.g., not equal to `0`). In addition, a feasibility analysis is performed in terms of the geographic location the measurement was conducted. For this purpose, we have created a database table with locally stored information about countries (e.g., name, continent, or population) from *GeoNames* including their boundary boxes and matched all countries to their corresponding MCCs. The class `ValidateCountry` uses this information and returns the expected MCC for the given geographic coordinates within the measurement. If the MCC matches the MCC received from the measurement, the insert operation into the database is performed.

The raw measurement data is inserted into the `openmobilenetwork` database consisting of tables starting with `app_`. An example of a complete measurement distributed in several database tables is given in Fig. 6.6.

The database `openmobilenetwork` consists of the tables `app_cells`, `app_cells_neighbors`, `app_cells_services`, `app_cells_traffic`, `app_gps`, `app_user`, and `app_wifi` that are related to each other via a unique `measurement_id` for storing raw network measurement data. For a particular measurement, `app_cells` serves as the main table that keeps track of the mobile network cell (and its attributes) the user was connected to. This data is complemented with the neighboring cells in `app_cells_neighbors` that were "seen" as well as the services (`app_cells_services`) used and the traffic generated (`app_cells_traffic`) on the smartphone during the measurement. WiFi access points, on the other hand, that were scanned within this particular measurement are stored in the `app_wifi` table. The location where the network data was collected as well as the specific time of the measurement are saved in the `app_gps` table with

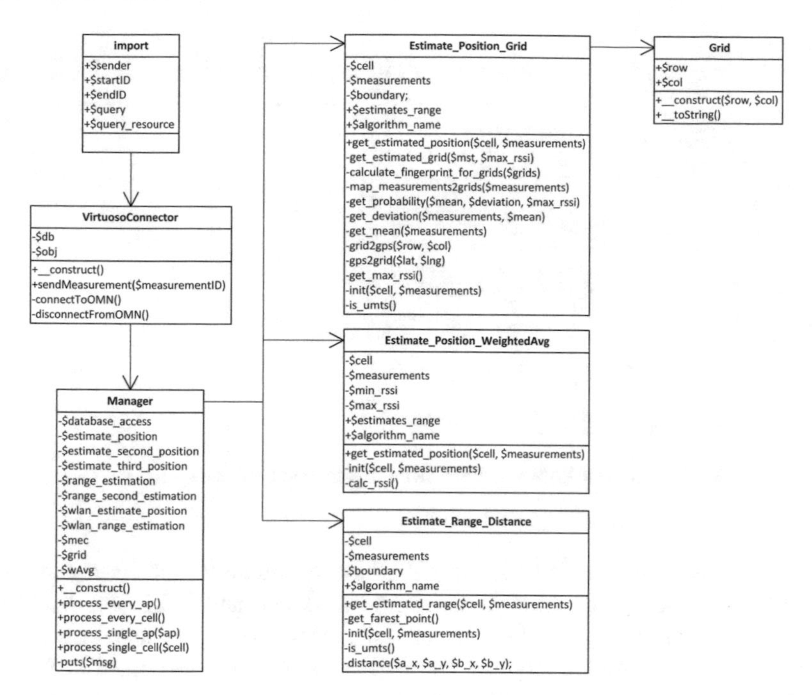

Fig. 6.7 OMN Measurement Data Manager – Position Estimation Manager

a relation to `app_user` through the `user_id`, which gathers information about the device and the SIM card of the crowdsourcing user. As soon as a complete measurement is stored into the database, the `measurement_id` of this measurement is written also into the `app_server_scripts` table for a later status check when approximating the position and coverage area of this cell.

After storing the network measurement into the relational database, the network topology estimation is started as part of the *Position Estimation Manager* service. In order to prevent the server from blocking due to too many processes, the *Position Estimation Manager* scripts are not running on the fly when receiving new measurements. A cronjob is set up to run `import.sh`, which again invokes `import.php` every 60 s in order to check if a semantification process (see Sect. 6.3.2) is running and, if not, if there are new measurements ready for network topology estimation. This status check is done by comparing whether the `measurement_id` of the last processed measurement in the `app_server_scripts` table is smaller than the `measurement_id` of the current measurement. If so, the position and coverage area is calculated (or updated) for this cell. This approach ensures sustainability and up-to-dateness of our dataset.

Figure 6.7 illustrates parts of the *Position Estimation Manager* service as a class diagram incorporating the functions for mobile network cell calculations. The `import` script creates an instance of `VirtuosoConnector`, which has two main functions. First, it triggers the *Position Estimation Manager* class (named as

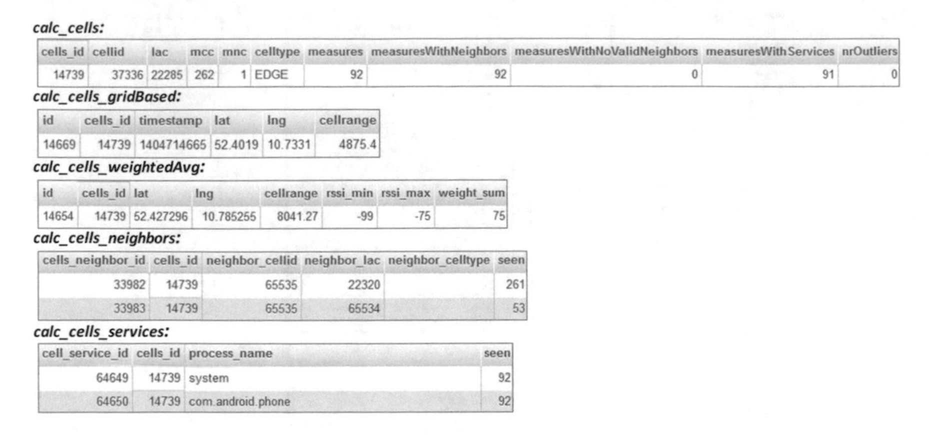

calc_cells:

cells_id	cellid	lac	mcc	mnc	celltype	measures	measuresWithNeighbors	measuresWithNoValidNeighbors	measuresWithServices	nrOutliers
14739	37336	22285	262	1	EDGE	92	92	0	91	0

calc_cells_gridBased:

id	cells_id	timestamp	lat	lng	cellrange
14669	14739	1404714665	52.4019	10.7331	4875.4

calc_cells_weightedAvg:

id	cells_id	lat	lng	cellrange	rssi_min	rssi_max	weight_sum
14654	14739	52.427296	10.785255	8041.27	-99	-75	75

calc_cells_neighbors:

cells_neighbor_id	cells_id	neighbor_cellid	neighbor_lac	neighbor_celltype	seen
33982	14739	65535	22320		261
33983	14739	65535	65534		53

calc_cells_services:

cell_service_id	cells_id	process_name	seen
64649	14739	system	92
64650	14739	com.android.phone	92

Fig. 6.8 OMN Measurement Data Manager – RDB calculated cell example

`Manager` within the implementation) in order to initiate the position and coverage area estimation process for the mobile network cell and WiFi APs included in the (new) measurement. For this purpose, the `Manager` class uses modularized functions for each network topology estimation algorithm as discussed in Sect. 4.4. The `Estimate_Position_Grid` class, for example, performs an approximation based on the grid-based approach, whereas the `Estimate_Position_WeightedAvg` class uses the weighted centroid-based algorithm for calculating the mobile network cell position. `Estimate_Range_Distance`, on the other hand, estimates the cell coverage area according to the design decisions in Sect. 4.4.6. We implemented several algorithms all of which are selectable for calculation based on the configuration of the service.

After successfully approximating (or updating) the network topology out of the network measurement, the data is written into tables beginning with `calc_` within the `openmobilenetwork` database. Figure 6.8 illustrates an exemplary dataset for a calculated cell. `calc_cells` is the main table and stores basic data about the mobile network cell and the number of measurements used for estimation, whereas the `calc_cells_gridBased` and `calc_cells_weightedAvg` tables keep track of the position and coverage area calculated by the respective algorithm. The neighboring cells that were measured by all measurements incorporating this specific cell are listed and counted within the `calc_cells_neighbors` table, while the services that were used during the collection of the measurements are comprised within the `calc_cells_services`. An equivalent set of database tables exists also for the calculated WiFi APs.

The second function of `VirtuosoConnector` triggers the *Semantification Manager* application for initiating the semantification process of the relational data. Details are given in the next section.

6.3.2 Semantification Manager

Two components are part of the *Semantification Manager* service. The *Network Data Semantification* function is responsible for "triplifying" relational data into RDF instances based on the *OpenMobileNetwork Ontology* and storing these statements into the Virtuoso triplestore, whereas the *LOD Cloud Interlinking* module serves for interlinking the semantically enriched data with external data sources of the *LOD Cloud*. Both components are developed as a Java application that is named `LiveDataVirtuoso`.

`LiveDataVirtuoso` provides two semantification modes. The *full semantification mode* is triggered manually and performs a semantification of the complete relational dataset if needed. This mode is initiated when collected data is already available in the relational database and the *Linked Data* platform is operated for the first time with an empty triplestore (or restored due to corruptness within the dataset). Here, the schema of the *OpenMobileNetwork Ontology* is initially imported into the triplestore before starting the semantification process. The *partial semantification mode*, on the other hand, is triggered by `VirtuosoConnector` as part of the *Position Estimation Manager* and stores RDF statements for new incoming cells or updates existing triples when attributes of calculated cells have changed due to new measurements.

Figure 6.9 shows a class diagram of the *partial semantification mode*, in which the `VirtuosoConnector` triggers the `LiveDataVirtuoso` application for the semantification process of the mobile network data. This application is separated into different modules, where each module defines a mapping schema for a specific network context facet of the *OMN Ontology* (see Sect. 4.5.1). The class `Cells`, for example, is responsible for triplifying the mobile network cell data based on the *Mobile Network Topology Ontology* facet, whereas the `NeighborCells` class converts the neighboring cells according to the *Neighbor Relations Ontology* facet. Transforming service usage data into its semantic representation given in the *Service Ontology* facet is handled by the `Services` class. The `Device` class, on the other hand, triplifies the mobile devices in use as modeled in the *Mobile Device Ontology* facet. Furthermore, the number of users as well as the traffic occurrences are calculated as described in Sect. 4.5.1.2 and converted by the `LiveUsers` and `LiveTraffic` modules according to the *Traffic and User Ontology* facet.

The *LOD Cloud Interlinking* function is incorporated into the semantification process and is initiated by the respective modules whenever needed. Geo-related datasets of the *LOD Cloud*, for example, are interlinked to the network topology data by creating an instance of the class `POI` within the `Cells` module and making use of its specific methods. Furthermore, the `Cells` class also instantiates an object of the class `StaticData` in order to interconnect the network operators and the countries they are located in, the network technologies as well as the mobile network generations to *DBpedia* resources describing them. Links to mobile device information within *DBpedia* are created directly within the `Device` class.

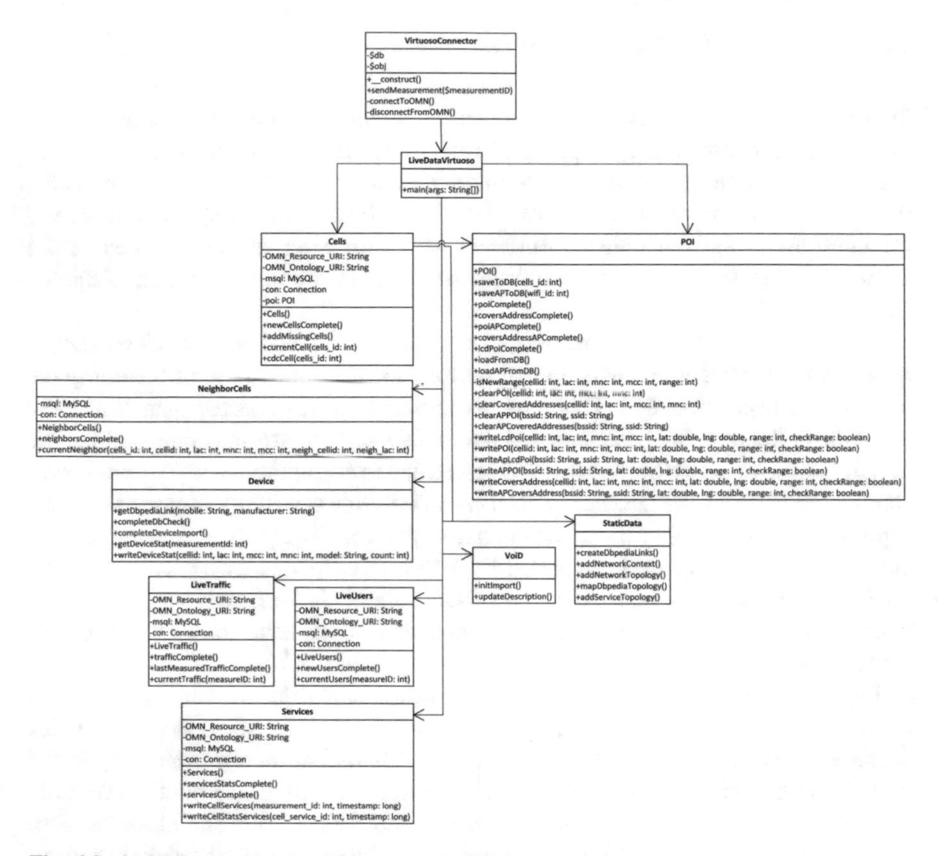

Fig. 6.9 OMN Semantification Manager – LiveDataVirtuoso

`LiveDataVirtuoso` serially processes all modules enabling a plug-in-like flexibility in the data to be semantified. Moreover, this modular approach is easily extendable making it adaptable to changes in the ontology and the dataset.

6.3.3 OpenMobileNetwork Website

The *OpenMobileNetwork* website comprises general information about the platform as well as the project and is responsible for publishing the dataset according to the principles of *Linked Data* by providing several representation forms of the information (see Sect. 6.3.3.1). The coverage maps (see Sects. 6.3.3.2 and 6.3.3.3), on the other hand, showcase various visualization options of the network topology data.

6.3.3.1 Published Linked Data

The dataset of the *OpenMobileNetwork* is published based on the principles of *Linked Data* and the best practices described in the tutorial of Heath and Bizer [62].

All resources that are available within our triplestore can be declared as "real-world objects" and are therefore non-information resources. They are identified by *HTTP Cool URIs* [131] and are dereferenceable using HTML as well as RDF browsers via content negotiation. Three representations are provided for each resource:

- http://www.openmobilenetwork.org/resource/cell2160591_5275_262_1, a URI identifying an exemplary non-information resource for the mobile network cell with the Cell-ID "2160591", the LAC "5275", the MCC "262", and the MNC "1",
- http://www.openmobilenetwork.org/data/cell2160591_5275_262_1, the RDF/XML representation of the resource,
- http://www.openmobilenetwork.org/page/cell2160591_5275_262_1, the HTML representation of the resource.

When designing our schema, we reused well-known vocabularies and complemented them with additional concepts needed for our dataset (see Sect. 4.5.1 for details). Furthermore, we paid attention to comment each concept (using `rdfs:comment`) for humans to understand, labeled them (with `rdfs:label`) appropriately, and stated all domains as well as ranges explicitly. In addition to the resources, all our ontology concept URIs are also dereferenceable, so that people can look up the meaning of them. The ontology URI for a mobile network cell, for instance, is defined as http://www.openmobilenetwork.org/ontology/Cell.

6.3.3.2 OpenMobileNetwork Coverage Map

The *OpenMobileNetwork Coverage Map*[12] displays worldwide network topologies using *OpenStreetMap* in combination with the *OpenLayers Map Viewer Library 2.5*,[13] *jQuery 1.11.1*,[14] and *jGrowl 1.2.6*.[15] In its current version, the weighted centroid-based algorithm is used for ordering the positions of the markers on the map. Figure 6.10 displays a map screenshot of the Schöneberg district in Berlin, where the *Telekom* office in the Winterfeldtstraße is located.

Here, the colored markers represent either a mobile network cell or a WiFi access point. For Germany, we have created an operator-specific visualization and separated each operator by a different color. Magenta stands for *Telekom*, whereas red represents *Vodafone*. *E-Plus* (now part of the O_2 brand) is visualized by green markers and O_2 by blue markers. Outside of Germany, all mobile network cells are represented by

[12]http://map.openmobilenetwork.org/.

[13]https://www.openlayers.org/.

[14]http://www.jquery.com/.

[15]https://plugins.jquery.com/jgrowl/.

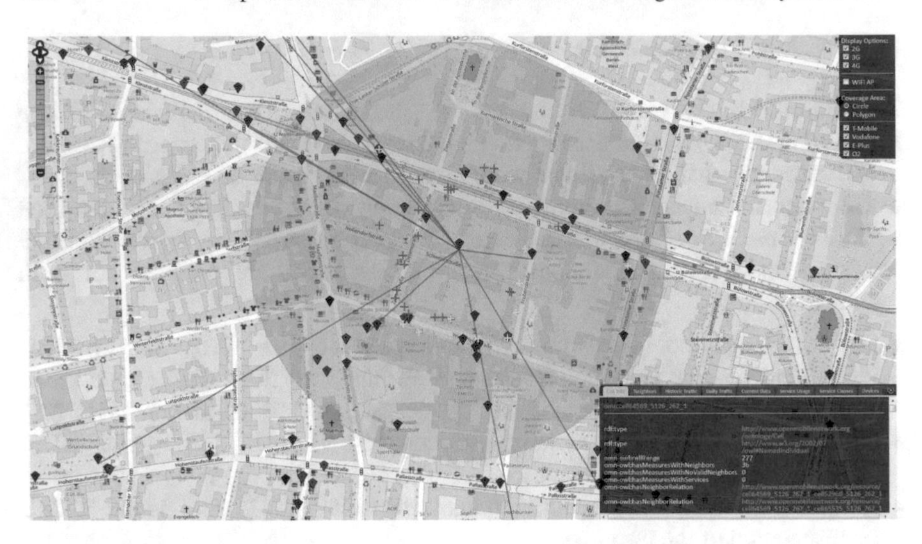

Fig. 6.10 OMN Coverage Map

white markers. Black markers, on the other hand, stand for WiFi APs, which are illustrated when checking the *WiFi AP* checkbox.

The *Coverage Area* display option enables a demonstration of the coverage as a circle or as a polygon for the selected marker. Other display options allow filtering the visualization by cell technology (e.g., 2G, 3G, or 4G) or by mobile network operators (only for Germany).

Clicking on one of the markers (e.g., mobile network cell or WiFi AP) shows the coverage area of the respective network component and opens the *Cell Information* tab that includes RDF data related to the cell or the WiFi AP such as the geo coordinates, the cell technology, or the radius of the coverage area. This tab also lists all links to location entities from *LinkedGeoData* as well as *Linked Crowdsourced Data* and mobile network operator information coming from *DBpedia*. Neighboring cell relations are listed in the *Neighbors* tab, which indicate in percentage how often a neighbor has been seen by smartphones when being connected to the certain cell. The *Traffic* tabs, on the other hand, show various traffic profiles created by the sum of all traffic measurements. In the *Historic Traffic* tab, traffic profiles for the last six months are visualized, whereas the last day on which traffic has been measured by a smartphone client is displayed in the *Daily Traffic* tab. The *Current Data* tab provides live traffic and user information collected at run-time by the smartphone clients. Service usage information is given in the *Service Usage* as well as *Service Classes* tabs and statistics about the mobile devices used within a mobile network cell are listed in the *Devices* tab.

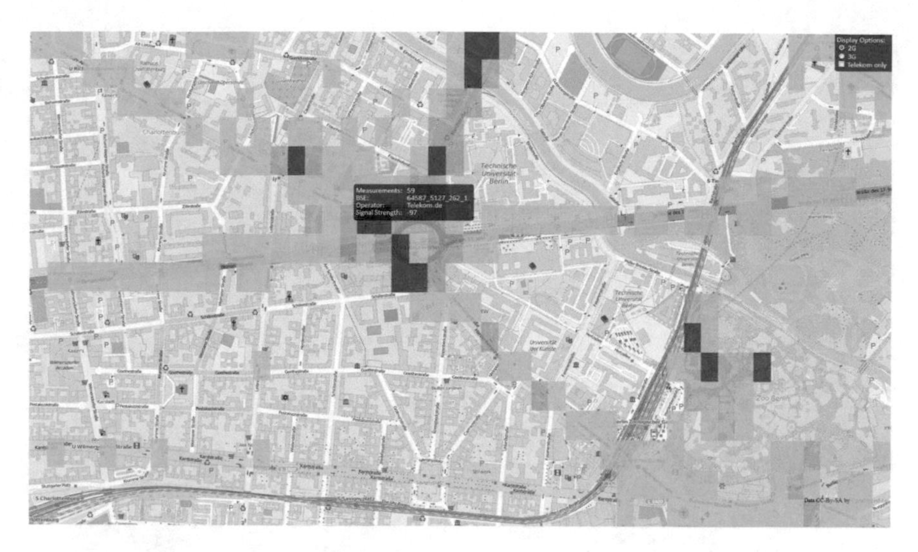

Fig. 6.11 OMN Best Server Estimates Map

6.3.3.3 OpenMobileNetwork Best Server Estimates Map

The *Best Server Estimates Map*[16] (see Fig. 6.11) uses the same technologies as mentioned in Sect. 6.3.3.2 and utilizes all single network measurements of the *OpenMobileNetwork* in order to identify the cells with the best connectivity for certain geographic areas.

These geographic areas are predefined as tiles of a dynamically calculated grid with 3×3 arcsecond per tile based on the shown map segment. Each network measurement is considered for calculating the number of measurements and the mobile network cell with the best signal strength per tile, so that in the end, each tile represents one cell with the best connectivity.

The grid is displayed as an overlay on the map. Tiles that contain no measurements are not displayed; others are grouped into three colors according to the number of collected measurements within each tile. Green represents less than 10 measurements, whereas orange stands for 10–49 measurements. The color red is used for 50 measurements and above.

A click on one of the tiles causes the grid to clear its features and to mark all other related tiles with the same cell where it performs best, i.e., where it is the cell with the highest signal strength. This view can be reset by clicking on a free map area or simply by dragging the map. Furthermore, a mouseover event forces a popup to come up displaying relevant cell information within a tile.

The visualization of the best server estimates is available for 2G as well as 3G cells and can be restricted to only use *Telekom* as an operator.

[16]http://www.openmobilenetwork.org/bse/map.php.

Chapter 7
Context-aware Services based on Semantically Enriched Mobile and WiFi Network Data

Chapter 7 introduces In-house, B2C, and B2B services that were designed and implemented as proofs of concepts for highlighting the added value of semantically enriched mobile and WiFi network data interlinked with other context data sources. While this chapter motivates the need for these services from a research perspective and describes the designed solutions, Chap. 8 gives an overview about the proof of concept implementations related to them.

In Sect. 7.1, we focus on an In-house service that enables operators to optimize their networks for power management purposes by de- and reactivating mobile network cells based on the capacity demand. Section 7.2, on the other hand, introduces several *Semantic Positioning* solutions that improve classic geocoding as well as geofencing methods and add semantic features to proactive self-referencing and cross-referencing LBSs. The usage of semantically enriched location analytics for B2B services is discussed in Sect. 7.3.

7.1 In-house Service: Power Management in Mobile Networks

Mobile network technologies, such as GSM, UMTS, or LTE, provide good connectivity and high data transfer rates, but always work in full capacity mode in order to fulfill the increasing demand on mobile network usage [1]. According to Biczók et al. [23], there were more than 4.6 million base stations deployed worldwide in 2013 and this number is expected to rise by 11.2 million in 2020. The permanent availability of those networks and the continuous infrastructure deployment leads to a significant energy consumption and CO_2 emissions. Here, 80–85% of the total energy consumption within a mobile network is caused by base stations [60].

A. Uzun, *Semantic Modeling and Enrichment of Mobile and WiFi Network Data*, T-Labs Series in Telecommunication Services,
https://doi.org/10.1007/978-3-319-90769-7_7

The provided mobile network capacities, however, are not always fully utilized or optimized to the real capacity demand. There is a huge network usage difference between day and night times, for example. But also geographic aspects, such as a city or a rural area, affect the network usage significantly. These different usage profiles enable the possibility to save energy in mobile networks through an adaptive and context-aware power management. Several solutions, such as a dynamic de- and reactivation of network components (e.g., base stations) or an adaptive network reconfiguration based on the user's needs, were developed within the *Communicate Green*[1] project.

Implementing such a complex context-aware power management for diverse network usage scenarios requires a correlation of various data sources such as network topology information (e.g., positions and coverage areas of mobile network cells) in combination with dynamic network and user context data including the amount of traffic generated in a mobile network cell, the number of users within a cell, their movement patterns, or service usage information. Moreover, information about POIs, events in certain areas, or weather conditions is of high relevance.

7.1.1 Network Optimization Use Cases

In order to highlight the requirements and demands in terms of network connectivity as well as QoE from a user perspective and to further identify the context data sources that are relevant for realizing a power management in mobile networks, we shortly describe a scenario of a typical user named John, who uses mobile connectivity in various situations of his daily life. Based on this scenario, several use cases are identified, which explain potential network optimizations triggered by context information that occur in these daily situations.

John works for an IT company, which is located in the business district of Berlin, Germany. He usually arrives at his office at 8 a.m. and leaves around 5 p.m. Due to the fact that with John there are thousands of other employees working in the same area during similar office hours, the mobile network capacity demand is very high during office hours and rather low at night times. After work, John usually spends some time in the city to run errands before going back home. With his smartphone, he takes pictures of the products that he wants to purchase in the city and sends them to his wife in order ensure that he selected the right ones. On the weekend, John loves to attend a game of his favorite soccer club in the *Olympic Stadium* with his friends. The stadium is very crowded during the soccer game; John and his friends are very excited, so that they record short videos of the action on the field and upload them to a video streaming portal. After the match, John, his friends, and the whole crowd leave the stadium, which leads to a great decrease of the network demand in that area.

The described scenario comprises two views on network optimizations. The microscopic view is user-oriented looking at a specific user - here exemplary named as John, the base station he is subscribed to, neighboring cells of the base stations and his end device. Optimizations are done based on this specific user and his surroundings.

[1] http://www.communicate-green.de/.

The *City Region* use case presented in Sect. 7.1.1.1 illustrates a microscopic view. The macroscopic perspective, on the other hand, is rather network-oriented and considers a big amount of users located in a certain area and mobility patterns created by them over time. An example for this view is the *Business Area and Stadium* use case in Sect. 7.1.1.2.

Section 8.1 presents the *OpenMobileNetwork for ComGreen Demo* that showcases and visualizes several power management use cases based on exemplary SPARQL queries.

7.1.1.1 City Region

In a regular city region with residential buildings, malls, and parks, user movements and network demands may differ on a daily or even hourly basis. Hence, real-time parameters of the region need to be taken into account when triggering network optimizations. One possible approach based on various types of context information could be as follows:

John wants to go shopping in the city before driving back home. At first, the geographic region of John is of importance since the network optimization algorithms and radio technologies to be used depend on the density of cells and the availability of different radio technologies in John's surroundings. From a user perspective, his location can be coarsely determined by his mobile device via the Cell-ID, which makes it possible to identify the distinct cell and area he is located in. Combining the information gained from his end device (e.g., the signal strength of neighboring cells, access technologies, and device capabilities) with a priori information of the network provider showing the cell distribution, cell types, cell technologies, and locations of the base stations, the cell topology of John's location in Berlin and his device's capabilities of accessing different radio technologies are identified. In the area, various radio technologies (e.g., GSM, UMTS, and LTE) may be deployed in the form of macro, micro, and pico cells, whereas an additional number of public and private WiFi hotspots is expected to be deployed in public cafés, shops, and offices as well as in private apartments. The exemplary average daily load (see Fig. 7.1) for the cell John is connected to, can be modeled as follows:

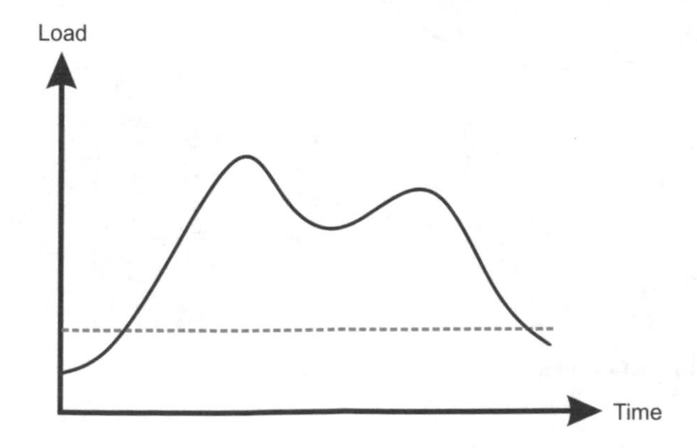

Fig. 7.1 City Region – exemplary average daily load

A load threshold for triggering network optimizations is defined based on a priori information. Due to the fact that the cell is located in the inner city and the cell topology in that area is very dense, the load threshold is defined much lower than in rural areas, for example. Furthermore, the average daily load is compared to the current load of the cell. If notable deviations from the expected load are detected, optimizations are triggered as well.

In a second step, the number of all users in the cell is used in order to further investigate if optimizations are possible. If the number of all users in the cell to be deactivated is below a predefined threshold and if cell coverage for different technologies as well as load and user capacity is given, forced handovers are performed. The forced handover decisions are based on a comparison of the services used by the remaining users in that cell using a service classification scheme, which lists what kind of service can be delivered by what kind of radio technology. The users are clustered based on this service classification. Depending on measurement reports (e.g., quality of the current link, RSSI of neighboring cells) and the actual capabilities of the mobile devices, users are handed over to an appropriate neighboring cell allowing the current cell to be deactivated when empty. Figure 7.2 illustrates the exemplary network optimization approach.

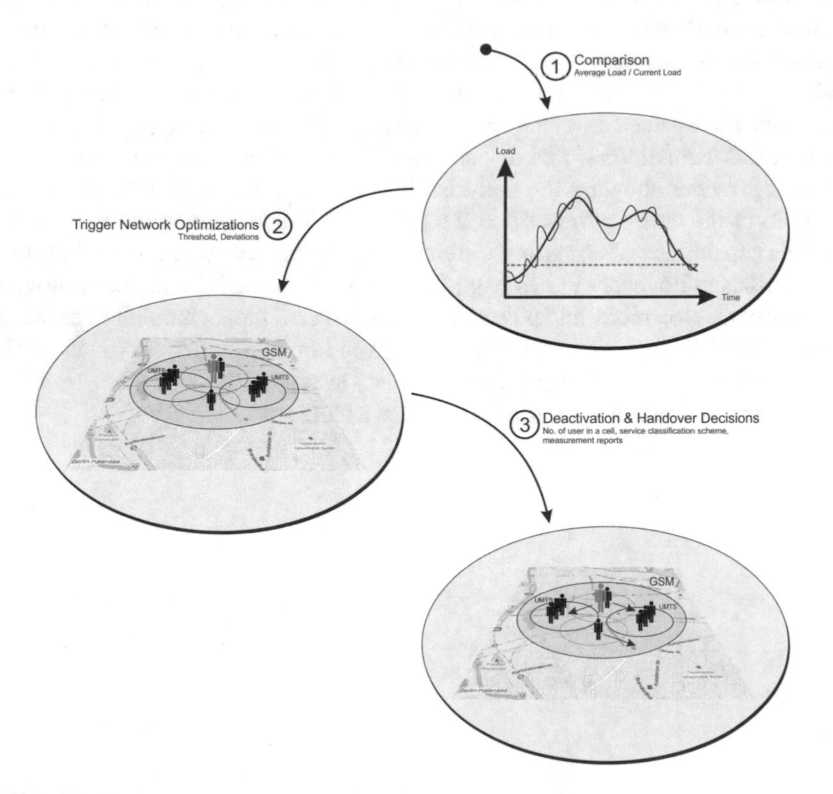

Fig. 7.2 City Region – exemplary network optimization approach

7.1.1.2 Business Area and Stadium

In order to fulfill the capacity demand in a business area or a place that hosts big events, such as a stadium, a network provider operates a large number of base stations of different access technologies. In addition, WiFi hotspots are installed at certain locations for complementing the provided mobile network bandwidth. By doing so, the network provider ensures that the expectations of the users in terms of mobile connectivity and bandwidth are satisfied during office hours or an event. However, a huge decrease of the capacity demand during night times in the business area or after the event in the stadium leads to an over-provisioning of network resources, which provides great potential for saving energy.

By applying prediction mechanisms in combination with valuable contextual information coming from external data sources, network elements that are not needed are put into sleep mode and the network is proactively reconfigured. Third-party services (e.g., Web APIs) or datasets within the *LOD Cloud* are used to retrieve data about the exact place and time of an event, the number of visitors and distribution of their age groups, for example. In addition, the websites of stadiums and other arenas usually include a location plan providing information about typical ways of entering and leaving people.

Mapping this use case to our current work, interlinking mobile network data provided by the *OpenMobileNetwork* with POIs from *LinkedGeoData* and soccer information available in *Linked Soccer Data* [16], for example, enables the operator to proactively optimize the network capacity needs in the area of the *Olympic Stadium* in Berlin during match times.

7.2 B2C Service: Semantic Positioning Solutions

Proactive LBSs form the next generation of LBSs, which persistently keep track of the users' locations in an unobtrusive manner and proactively send a notification about potentially useful information in the vicinity according to presubscribed events [10]. They make use of spatial geofences based on geometric shapes (e.g., circles or polygons) that are defined around all relevant location entities (e.g., POIs or events) in order to perform an action as soon as a user enters or leaves (one of) the geofences. These services are further distinguished between self-referencing and cross-referencing dependent on whether the user and the tracked target is identical or different [84].

Conventional (proactive) LBSs suffer from a number of drawbacks: First, they require manual effort in defining geofences beforehand around all relevant location entities by utilizing *Geofencing APIs* or visual editors. For example, if a well-known fast food company (e.g., *McDonald's*) provided an LBS that offers its customers vouchers as soon as they enter one of their restaurants, the application developer would need to manually define geofences around all *McDonald's* worldwide, which is obviously time-consuming. In addition, a permanent comparison in terms of the intersection between a user's position and a geofence (defined as a polygon) is computationally expensive [70].

Second, LBSs are solely driven by geometric spatial data. However, the information corresponding to a location is much more than the abstracted geographic coordinates. More fine-grained semantic knowledge about the location, such as near POIs or landmarks, user-specific locations (e.g., home, office, school, etc.), events occurring in the vicinity, or its relation to other neighboring locations, is a valuable information source for LBS applications, especially when providing personalized features where two users submitting the same request at the same location and time are expected to receive different notifications according to their user profile. Taking a basic *restaurant recommender* service as a proactive LBS example, geofences are defined around restaurant locations. A user is notified about a restaurant upon entering one of the defined geofences. Further semantic information about the location and the user (e.g., "a restaurant serving a user's specified favorite meals", "a restaurant within a user's defined budget", or "a restaurant that people like to go at night during summer") is simply not considered [C9].

A third challenge is related to geocoding services, which strongly depend on the correctness and completeness of the source and input address data. Incomplete or ambiguous input data, such as a street name without the postal code that exists several times in a city, often leads to wrong results in the geocoding process. In [157], the authors show that different geocoding systems tend to yield different results for those hard cases.

In this section, we highlight how semantically enriched mobile network and WiFi AP topology data can be leveraged to tackle the above mentioned drawbacks for enabling a better LBS experience. For this purpose, we present several *Semantic Positioning* solutions relying on the *OpenMobileNetwork* that improve classic geocoding as well as geofencing methods and add semantic features to proactive self-referencing and cross-referencing LBSs. Our contributions comprise

1. a *Semantic Tracking* approach that provides geofencing and continuous background tracking functionality for self-referencing as well as cross-referencing LBSs without the need to create geofences for environments densely covered by networks (e.g., cities). This is done by geographically and semantically relating mobile network and WiFi AP topology data to other geo-related datasets (e.g., LGD or LCD [C6]) in order to map POIs, for example, to the coverage areas of the network infrastructures. These semantic relations also avoid a complex *point-in-polygon* comparison decreasing computational costs,
2. an integration of semantic information into the *Semantic Tracking* process such as favorite or frequently visited locations as well as popular places within a certain contextual situation,
3. a *Semantic Geocoding* that interlinks the network data with structured address data gained from OSM in order to improve mobile geocoding services in cases of ambiguous and incomplete address queries.

7.2.1 Semantic Tracking

Conventional proactive LBSs perform tracking with respect to manually created geofences and a matching of WGS84 coordinates to those geofences. Our proposed *Semantic Tracking* [C9, C10] approach does not rely on geofences nor geo coordinates. It fully exploits semantic and geographic relations between wireless network topologies, location entities being in the coverage areas of those networks and semantic user profiles. Enabling self-referencing as well as cross-referencing service features is based on three fundamental requirements:

1. Semantically structured mobile network and WiFi access point topology data (including their coverage areas) mapped onto geographic regions
2. Knowledge about a user's profile including preferences, social relations to other people, favorited location entities (e.g., the user likes to spend time in parks during summer), and places where he is present frequently (e.g., school or office)
3. Tracking options (e.g., notifications upon entry or exit of a target user) and permissions

As described in Sect. 4.5.1, the first aspect is considered within the *OMN Ontology* in combination with interlinks to other geo-related datasets. Knowledge about a user and his preferences, on the other hand, is modeled within the *CDC Ontology* that consists of a number of ontology facets (see Sect. 5.2.3). By combining concepts from the *User Profile*, *Location*, as well as *Tracking and Service* facets, diverse aspects, such as basic user information, social relations, (location) preferences, service usage information, or tracking options, are supported.

Third-party tracking is the third fundamental prerequisite for enabling cross-referencing LBSs. These services need certain requirements to be fulfilled in order to ensure privacy and control over the tracking process. First, the social relationship between two users is of relevance since tracking can only be performed among people that know each other. The stronger the relationship (e.g., acquaintance, good friend, or family), the more permissions for tracking can be given. Secondly, the tracked target must always have the possibility to permit or forbid tracking even if he granted tracking rights to a person. Furthermore, the tracking user should have options to activate and deactivate tracking or to adjust the notification options (e.g., upon entry or exit of the target user).

The *Tracking* facet as part of the *CDC Ontology* defines ontology concepts that incorporate these requirements. Using this ontology facet, tracking is achieved by utilizing the semantic relations between location entities linked to mobile and WiFi networks, social relationships between people, and tracking options as well as permissions defined between them.

Proactive self-referencing and cross-referencing LBSs involve continuous tracking of users' locations in the background for detecting whenever a presubscribed context-event has been fulfilled. The *Semantic Tracking* approach solely relies on the network information retrieved from the mobile client (i.e., a base station's Cell-ID or the BSSID of a WiFi AP), which increases its energy efficiency and suffices

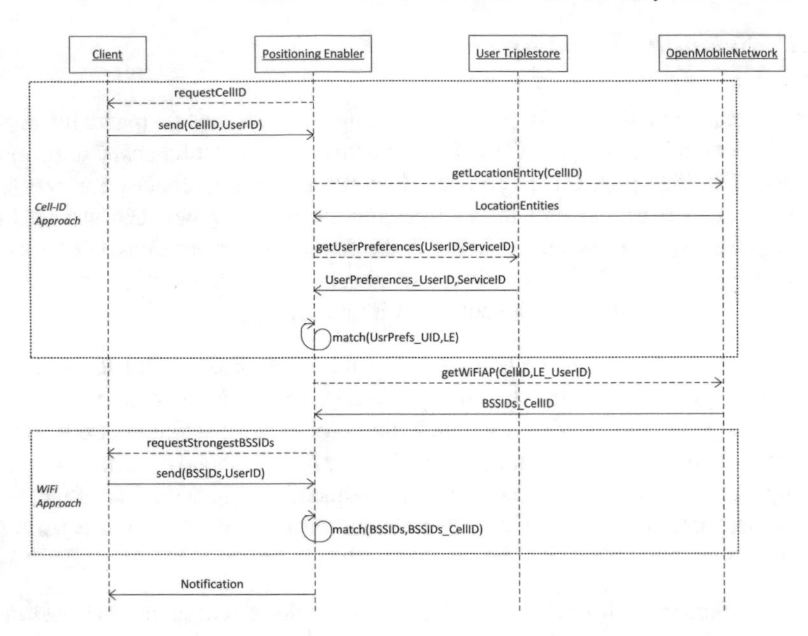

Fig. 7.3 Semantic Tracking – background tracking strategy

for most of the proactive LBS scenarios in terms of tracking accuracy. It further takes user profile information into consideration for tracking the users and therefore promotes the delivery of highly personalized services.

For realizing the background tracking service, the *Positioning Enabler (PE)* function of the CDC platform requests the mobile client to send network information that comprises either the Cell-ID of the base station to which the mobile client is connected to or the BSSIDs of the WiFi APs with the strongest received signal strengths. In order to enable continuous tracking, implementing a background service on the mobile client is inevitable, which periodically provides the server with the relevant information required for tracking the users.

The logic behind the *Semantic Tracking* approach is illustrated in Fig. 7.3. Upon creation of a user profile and subscribing for an LBS, the PE turns the background service on the mobile client on and starts tracking the user. It starts with the *Cell-ID Approach* requesting the mobile client to send the Cell-ID of the base station that it is connected to, as well as the *UserID* denoting the user identity. The PE resolves the Cell-ID using the OMN and gets all location entities (referred to as `cdc-owl:LocationEntity` in Fig. 4.11) covered by the coverage area of that cell. It then asks for the preferences of the user from the *User Triplestore* (see footnote 14 in Chap. 5). These could either be the locations a user has predefined in his user profile as favorite or frequently visited location entities or any other preferences (e.g., restaurants that belong to a certain cuisine). Afterwards, the PE checks if any of the location entities contained within the Cell-ID the user is present in, are of interest to him based on the preferences in his profile. If the PE finds no interesting location entities for the user, no action is taken and the steps of the *Cell-ID Approach* are repeated when entering a different cell. Under the condition that the PE finds interesting location entities within the user's cell, it requests the OMN to provide the BSSIDs of the WiFi APs in the cell (see Fig. 4.14), which cover those location entities. If the OMN returns no WiFi BSSIDs, then this implies that the interesting location entities within that cell are not covered by WiFi APs and hence a notification is sent to the tracking user notifying him of a location event, which has a cell-wide accuracy.

In case the OMN returns WiFi BSSIDs, the PE then switches to the *WiFi Approach* requesting the mobile client to send the (three) BSSIDs of the WiFi APs with the strongest received signal strengths, hence strives to position the user more accurately. It then compares the user's sent BSSIDs with the WiFi BSSIDs containing interesting location entities (which was provided earlier by the OMN). If the BSSIDs match, then the tracking user is notified of a location event in his vicinity. If no BSSIDs match, then the steps of the *WiFi Approach* are repeated every 10 s until either a match is found or the user exits the cell and the *Cell-ID Approach* starts all over.

We have integrated the *Semantic Tracking* functionality into the *CDCApp* in the form of a *Popular Places Finder* and *Friend Tracker* service in which a user gets notified about popular places in his vicinity and whenever his friends leave or enter a favorite or frequently visited location such as the office. Exemplary SPARQL queries for these services are given in Sect. 8.2.

7.2.2 *Semantic Geocoding*

Geocoding is the process of deriving WGS84 coordinates for a location specified in a human-readable format. Nowadays, geocoding services are often used via smartphone. This mobile geocoding use case comprises slightly different challenges and introduces more complexity when compared to generic solutions. While a general query could have referred to any address in the world, the user is most probably looking for a destination nearby when using his mobile device. In addition, typos are more likely to happen and the input is more likely to be ambiguous due to the fact that the user might specify a nearby address incompletely by omitting administrative areas that are obvious in his context.

The *Semantic Geocoding* approach targets this mobile geocoding use case and combines a geocoding query with the user's location context (sensed from his smartphone) in order to derive more accurate results even if the query itself is less precise. Here, the location context incorporates information about the mobile network cell the user is connected to, e.g., the Cell-ID, the LAC as well as the MCC. The MCC acts as an indicator for the country in which the target address is most likely to be found as well as for the format of the address itself. In contrast to this broad indicator, the Cell-ID is used for locating addresses in the coverage area of the cell. Additionally, the LAC supports the search for addresses in nearby cells of the same mobile network operator that most probably share the same LA. Having this context data available in varying precision and hence knowing in "which part of the town" the user is located, allows the *Semantic Geocoding* to determine which address the user most probably has referred to in case of an ambiguous query.

Our algorithm for the *Semantic Geocoding* is based on the *OMNG Ontology* and the geographic mapping of address data to the mobile network topology information of the *OpenMobileNetwork* (see Sect. 5.3). At first, the algorithm utilizes a set of regular expressions to parse the input address according to address formats used in the country given by the MCC. Once the individual components have been identified, the search begins by checking whether the address is among those that are covered by the given Cell-ID using the `omng-owl:isCoveredByCell` property. If no results are delivered, the search is extended to the entire LA with the `omng-owl:isCoveredByLAC` property considering all nearby cells having the same LAC. In case this step returns no results either, the location context information is removed and the entire dataset is crawled. At this stage, the geocoder is comparable to existing generic solutions, i.e., no additional location context information is used to enhance the search. If this stage also yields no results, the procedure is repeated from the beginning, gradually removing the address parts *city*, *postal code*, and *house number*. This step-by-step process leads to an unrestricted global street name search by the time of the last query. Therefore, each query is less precise than the previous one and missing data as well as erroneous input are ruled out successively. If multiple results are found at any stage, the first one is always chosen. However, due to the fact that we utilize location context information in the first place, ambiguities and thus the scenario of multiple results is avoided in most cases.

A demonstration of the *Semantic Geocoding* approach is shown in Sect. 8.3 in the form of the *OpenMobileNetwork Geocoder*.

7.3 B2B Service: Semantically Enriched Location Analytics

As discussed in [C6, C11] and Sect. 5.2, mobile network data is a valuable asset that can be used for establishing network operators as service enablers who become capable of providing B2B services based on location analytics. *Telefónica Smart Steps* [15], for example, is such a product that analyzes crowd movements out of mobile network data for providing footfall information to businesses for a specific location. *MotionLogic* (see footnote 4 in Chap. 1), a startup belonging to *Deutsche Telekom*, also utilizes traffic and movement streams for offering "geomarketing insights" to business customers.

Nevertheless, the exploitation of mobile network data for location analytics services is still restricted due to the fact that none of these approaches consider enriching their services with semantic information. Furthermore, even though network providers are able to infer highly frequented locations based on user movements, the results are mostly given on an abstract geographic level without having knowledge about the exact shop on a strip mall, which is highly visited under specific contextual circumstances, for example.

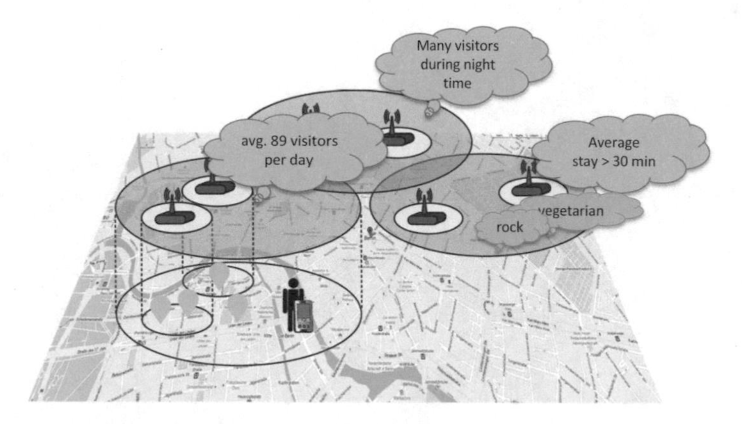

Fig. 7.4 Location analytics based on network topology data

These challenges are addressed by utilizing the semantically enriched mobile and WiFi network data within the *OpenMobileNetwork* in combination with the dataset of *Linked Crowdsourced Data* that extends static location-related data with diverse and dynamic context information. By doing so, information about the "popularity" of certain places or the "visiting frequency" of users in specific contextual situations (e.g., during certain weather conditions or on holidays) is made available in *Linked Data* format as part of the *LOD Cloud*.

Linking the *OpenMobileNetwork* to *Linked Crowdsourced Data* enables a contextual view on mobile network data and hence allows network operators to offer fine-grained and semantically enriched location analytics services in a B2B relationship. Figure 7.4 illustrates a visual example where telecommunication providers become capable of formulating sophisticated SPARQL queries in order to retrieve highly personalized and contextual information for their own operated cells, e.g., "the average number of users being connected to a certain cell when its rainy outside" or "the number of cells consisting of mostly vegetarian people that prefer dishes under 10 euros".

Aggregated and rudimentally analyzed LCD data in terms of location analytics is visualized on the *Location Analytics Map* introduced in Sect. 8.4. A location analytics application scenario, on the other hand, is given in Sect. 8.2.2 in the form of a *Popular Places Finder* service.

Chapter 8
Service Demonstrators

In conjunction with Chap. 7, this part gives an overview about our proof of concept demonstrators ranging from In-house to B2C services that utilize semantically enriched mobile and WiFi network data in combination with other context data sources. Section 8.1 introduces the *OpenMobileNetwork for ComGreen Demo* as an In-house network service [C11] including exemplary SPARQL queries that showcase several power management use cases in which the network operator is looking for candidate cells to be deactivated in order to save energy. Section 8.2, on the other hand, gives background tracking query examples for the *Friend Tracker* and *Popular Places Finder* services of the *CDCApp* that are based on our *Semantic Tracking* (see Sect. 7.2.1). In Sect. 8.3, the Web interface of the *OpenMobileNetwork Geocoder* is presented, which enables a direct map comparison of address query results between our *Semantic Geocoding*, *Nominatim*, and *Google*, whereas the *Location Analytics Map* in Sect. 8.4 [C6] visualizes aggregated and rudimentally analyzed LCD data.

8.1 OpenMobileNetwork for ComGreen

As an In-house network service example, the *OpenMobileNetwork for ComGreen Demo* showcases several power management use cases in which the network operator is looking for candidate cells to be deactivated in order to save energy. For this purpose, we have enhanced the standard *OpenMobileNetwork Coverage Map* by a number of specific checkboxes that run sophisticated SPARQL queries in the background.

The use cases differ in the variety of interlinked context data sources. Section 8.1.1 describes a basic and obvious power management use case incorporating the network topology, POIs as well as traffic information, whereas Sect. 8.1.2 extends the power management by also taking service usage information of the end devices into consid-

A. Uzun, *Semantic Modeling and Enrichment of Mobile and WiFi Network Data*, T-Labs Series in Telecommunication Services,
https://doi.org/10.1007/978-3-319-90769-7_8

eration. Section 8.1.3, on the other hand, shows a general example of how to aggregate geographically restricted traffic and user data over the network in order to analyze recurring patterns and work on prediction mechanisms for power management.

8.1.1 Use Case 1: Identification of Candidate Cells

Utilizing the relations between the serving mobile network cell and its neighbors, a click on the *ComGreen* checkbox on the map identifies cells covering a specific POI that are candidates to be deactivated in a power management process and further visualizes all potential neighboring cells for users to be handed over after a deactivation. In our POI example, this is done by querying the triplestore for all cells covering *T-Labs* at Winterfeldtstraße (`lgd:way33557476`) in Berlin, Germany. These cells need to have at least one neighboring cell with a connection availability of more than 60% as shown in Listing 8.1.

```
PREFIX omn:<http://www.openmobilenetwork.org/resource/>
PREFIX omn-owl:<http://www.openmobilenetwork.org/ontology/>
SELECT DISTINCT ?cell, ?lat, ?lng, ?cellid, ?lac, ?neighLat, ?neighLng,
                ?time, ?rx, ?tx
{
  ?cell rdf:type omn-owl:Cell .
  ?cell geo:lat ?lat .
  ?cell geo:long ?lng .
  ?cell omn-owl:hasMeasuresWithNeighbors ?number .
  ?cell omn-owl:operatedBy omn:mnc262_1 .
  ?cell omn-owl:hasNeighborRelation ?neigh .
  ?cell omn-owl:hasCellID ?cellid .
  ?cell omn-owl:hasLAC ?lac .
  ?neigh omn-owl:hasBeenSeen ?seen .
  ?neigh omn-owl:hasNeighbor ?neighcell .
  ?neighcell geo:lat ?neighLat .
  ?neighcell geo:long ?neighLng .
  ?cell omn-owl:hasTraffic ?traffic .
  ?traffic omn-owl:hasTrafficEvent ?traffic_event .
  ?traffic_event omn-owl:hasRxTrafficPerMin ?rx .
  ?traffic_event omn-owl:hasTxTrafficPerMin ?tx .
  ?traffic_event omn-owl:hasUniqueTime ?utime .
  ?utime rdfs:value ?time .
  ?cell omn-owl:covers <http://linkedgeodata.org/triplify/way33557476> .
  FILTER(?number>0 && (xsd:double(?seen)/xsd:double(?number))>0.6)
}
ORDER BY DESC[?time]
```

Listing 8.1 SPARQL query for identifying candidate cells

The underlying script checks if the returned traffic values represent live data. If so, a limit of 50 KB/min each for received and transmitted traffic is set in order to filter cells where traffic currently occurs. If the latest traffic data is older, a low traffic consumption is assumed. In this case, the filter is not applied and the mobile network cells as well as their neighbor relations are returned immediately. The candidate cells to be deactivated are displayed by yellow markers and the potential neighbors for a handover process are visualized by yellow lines to the neighboring cells.

8.1.2 Use Case 2: Service Usage Statistics

In the *OpenMobileNetwork*, service usage information includes statistical information as well as the traffic produced by each service on the end device. For each cell, a counter function computes how often a service has been used and how much traffic it has produced in total, so that we become capable of generating service usage statistics. These statistics comprise historic usage information calculated out of the total traffic ever generated by each service in the past as well as daily usage data created for the last day in which service measurements were made.

To demonstrate the applicability of service usage information for a power management in mobile networks, we have visualized the services as well as the service classes on the *OpenMobileNetwork Coverage Map* (see Fig. 8.1). The *Service Usage* checkbox fires the SPARQL query given in Listing 8.2 retrieving all mobile network cells covering *T-Labs* at Ernst-Reuter-Platz (`lgd:way40452037`) in Berlin and having service information on `omn-owl:MediaStreaming`, which is described as the ratio of the traffic produced by `omn-owl:MediaStreaming` services to the total traffic of the mobile network cell.

The underlying script filters the given information for the last day where service information is available. The expression `(xsd:double(?rx+?tx)/xsd:double(?rx2+?tx2))` describes the ratio of the traffic consumption produced by `omn-owl:MediaStreaming` to the total traffic of the mobile network cell. Cells with a value less than or equal to `0.4` are filtered and demonstrated on the map by yellow markers.

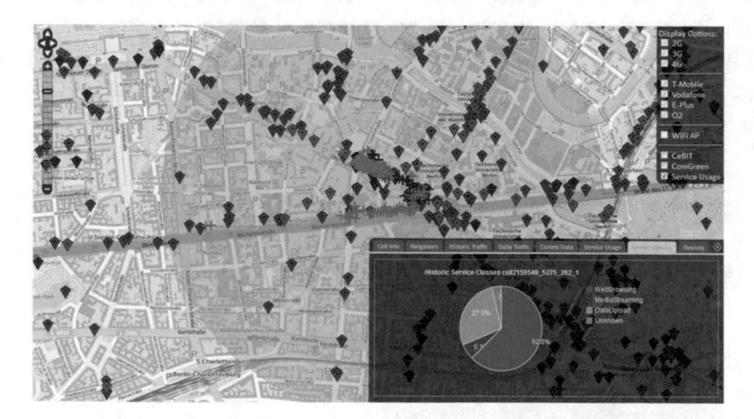

Fig. 8.1 OpenMobileNetwork for ComGreen Demo - service usage statistics

```
SELECT ?lat, ?lng, ?cellid, ?lac, ?time,
       (xsd:double(?rx+?tx)/xsd:double(?rx2+?tx2))
{
  ?cell rdf:type omn-owl:Cell .
  ?cell geo:lat ?lat .
  ?cell geo:long ?lng .
  ?cell omn-owl:operatedBy omn:mnc262_1 .
  ?cell omn-owl:hasCellID ?cellid .
  ?cell omn-owl:hasLAC ?lac .
  ?cell omn-owl:hasServiceEvent ?serviceEvent .
  ?serviceEvent omn-owl:hasService ?service .
  ?service rdf:type omn-owl:MediaStreaming .
  ?serviceEvent omn-owl:hasTrafficEvent ?trafficEvent .
  ?trafficEvent omn-owl:hasRxTrafficPerMin ?rx .
  ?trafficEvent omn-owl:hasTxTrafficPerMin ?tx .
  ?cell omn-owl:hasServiceEvent ?serviceEvent2 .
  ?serviceEvent2 omn-owl:hasTrafficEvent ?trafficEvent2 .
  ?trafficEvent2 omn-owl:hasRxTrafficPerMin ?rx2 .
  ?trafficEvent2 omn-owl:hasTxTrafficPerMin ?tx2 .
  ?trafficEvent omn-owl:hasUniqueTime ?uniqueTime .
  ?uniqueTime rdfs:value ?time .
  ?cell omn-owl:covers <http://linkedgeodata.org/triplify/way40452037>
}
ORDER BY DESC[?time]
```

Listing 8.2 SPARQL query for retrieving cells with specific service usage information

8.1.3 Use Case 3: Traffic and User Calculation

Use Case 3 shows a general example of how to aggregate geographically restricted traffic and user data over the network in order to analyze recurring patterns and work on prediction mechanisms for power management. For this purpose, the SPARQL query given in Listing 8.3 queries the *OpenMobileNetwork* triplestore for all mobile network cells located in and around Berlin, Germany, in terms of traffic and user information.

```
SELECT ?cell ?lat ?long (SUM(?traffic) AS ?totalTraffic)
       (xsd:integer(?number)/xsd:integer(COUNT(DISTINCT ?event))
       AS ?totalUniqueUsers)
WHERE {
  ?cell geo:long ?long .
  ?cell geo:lat ?lat .
  ?cell omn-owl:hasServiceEvent ?service .
  ?service omn-owl:hasTrafficEvent ?trafficEvent .
  ?trafficEvent omn-owl:hasRxTrafficPerMin ?traffic .
  ?cell omn-owl:hasUserHistory ?userHistory .
  ?userHistory omn-owl:hasUserEvent ?event .
  ?event omn-owl:noUserPerMin ?number .
  ?event omn-owl:hasUniqueTime ?time .
  ?time rdfs:value ?trafficTime .
  FILTER(bif:st_distance((bif:st_point(?long, ?lat)),
        (bif:st_point(13.38, 52.51))) < 100)
  FILTER(?trafficTime > 1349049600)
  FILTER(?trafficTime < 1351641600)
}
ORDER BY DESC(?totalUniqueUsers)
```

Listing 8.3 SPARQL query for traffic and user calculation

Table 8.1 Excerpt of the results for the SPARQL query in Listing 8.3

cell	lat	long	totalTraffic	totalUniqueUsers
omn:cell563_20453_262_3	52.538067	13.327541	13409	4
omn:cell132637641_40023_262_3	52.538826	13.346797	158722	3
omn:cell15693_20453_262_3	52.521622	13.352103	228914	2
omn:cell132616961_40023_262_3	52.518276	13.345784	371423	2

The filtering according to the location is applied using the `bif:st_distance` function. Moreover, for each cell, its approximated location is returned as geographic coordinates (using `geo:lat` respectively `geo:long` as seen in the ontology in Fig. 4.11) as well as the number of connected users and the traffic generated (in Bytes) by those users. The result is ordered according to the number of users connected to the corresponding cell. The `FILTER` statements define an interval specified in POSIX time for which the data is retrieved. In the given example, the data is aggregated over the entire month of November, 2013.

Table 8.1 presents an excerpt of the results for the SPARQL query in Listing 8.3. This information can support the network provider to decide, whether a certain cell can be deactivated. Instead of ad-hoc decision making, a viable approach is to analyze the underlying data for recurring patterns, so that the network operators can predict the load of cells. This can be achieved by iteratively executing the above-mentioned query while adjusting the time interval for which the data is retrieved. In doing so, peaks and low points for traffic and connected users can be estimated and put into perspective. As an example, on weekends in the summer time, the cells covering popular bathing lakes might be much more crowded than usual, whereas during Christmas time, cells covering the big Christmas markets should have a larger number of connected users.

In [C2] and [C4], we follow this approach and present the *OpenMobileNetwork Predictions Visualizer*, which demonstrates several prediction scenarios including traffic and user movement predictions based on the mobile network data of the *OpenMobileNetwork*. However, this work if out of scope within this thesis.

8.2 Semantic Tracking Services of the CDCApp

As described in Sect. 6.2.3, the *CDCApp* offers a set of context-aware services to users, such as a *Friend Tracker* [C10] or a *Popular Places Finder*, that utilize our *Semantic Tracking* approach for self-referencing as well as cross-referencing LBSs.

This section showcases the background tracking process by assembling a set of exemplary SPARQL queries used within these services.

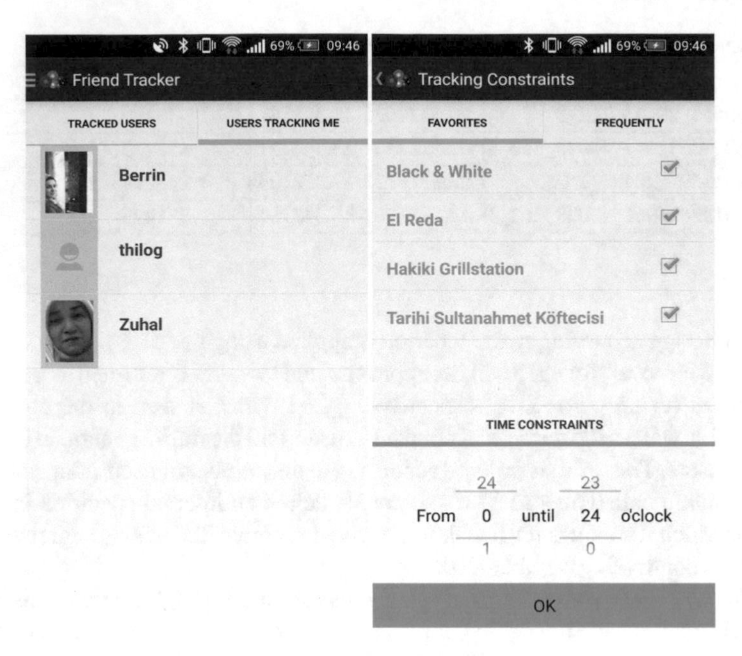

Fig. 8.2 CDCApp – Friend Tracker Service

8.2.1 *Friend Tracker*

By using the *Friend Tracker* service of the *CDCApp*, friends and family members are able to track each other according to their favorite and frequently visited locations. The service continuously tracks users in the background by utilizing the semantically enriched mobile network and WiFi access point topology data of the *OpenMobileNetwork* in combination with other (geo-related) data and user profile information (see Sect. 7.2.1). Users get notified whenever their friends or family members leave or enter a favorite or frequently visited location entity such as their college or school. Figure 8.2 presents screenshots of this service.

Our SPARQL query example for the *Friend Tracker* service demonstrates the *Semantic Tracking* approach for cross-referencing LBSs. For this purpose, we assume that there is a `cdc:user00002` of the CDC platform who is registered for the service, added `cdc:user00001` as his friend and asked him for tracking his office (`cdc:loc183` representing *Telekom Innovation Laboratories*) calculated as a frequently visited location. We further assume that `cdc:user00001` is currently connected to the mobile network cell `omn:cell2109543_5275_262_1`.

As demonstrated in Listing 8.4, the tracking process starts with a query for retrieving all location entities covering this cell from the SPARQL endpoint of the OMN. The result is a set of location entities, which also comprises `cdc:loc183` as being *Telekom Innovation Laboratories*.

```
PREFIX omn-owl: <http://www.openmobilenetwork.org/ontology/>
PREFIX omn: <http://www.openmobilenetwork.org/resource/>
SELECT ?loc {
  omn:cell2109543_5275_262_1 omn-owl:coversLocationEntity ?loc
}
```

Listing 8.4 Querying *sparql.openmobilenetwork.org* for location entities covered by a certain cell

Before `cdc:user00002` is able to track `cdc:user00001`, the system needs to make sure that `cdc:user00001` is registered for the *Friend Tracker* service. This is checked by a query to the *User Triplestore* (see Listing 8.5). The result of this query is the `?serviceName` ("Friend Tracker") and the `?id` (equal to `1`) of the service.

```
PREFIX cdc-owl: <http://www.contextdatacloud.org/ontology/>
PREFIX cdc: <http://www.contextdatacloud.org/resource/>
SELECT ?serviceName ?id WHERE {
  cdc:user00001 cdc-owl:usesService ?service .
  ?service cdc-owl:hasServiceName ?serviceName .
  ?service cdc-owl:hasServiceId ?id .
}
```

Listing 8.5 Querying *contextdatacloud.org:8891/sparql-auth* for the services of `cdc:user00001`

The query in Listing 8.6 demonstrates the matching of the location entities found in Listing 8.4 to the tracking permissions and constraints of the tracking user(s). All `?trackerUsers` are retrieved that have a tracking relation (`?tr`) to the user `cdc:user00001` with a valid `cdc-owl:TrackingPermission` (equal to `1`). This tracking relation needs to have an active (`cdc-owl:activeTracking` equal to `1`) location entity to be tracked (`?trackLoc`) that `cdc-owl:refersTo` a frequently visited location (`?freqLoc`), which again `cdc-owl:refersTo` to one of the location entities found in Listing 8.4 (e.g., `cdc:loc183`). The time constraint filter checks whether the time interval (e.g., `cdc-owl:intervalFrom` equal to 08:00:00 and `cdc-owl:intervalTo` equal to 20:00:00) in which `cdc:user00001` allows to be tracked matches the current time of the `?trackerUsers`. With the `cdc-owl:trackingNotificationOption` filter, it is checked whether it is an *entering* (equal to `1`) or *on-change* (equal to `3`) event.

```
SELECT *
{
  ?trackerUser cdc-owl:hasTrackingRelation ?tr .
  ?tr cdc-owl:tracksUser cdc:user00001 .
  ?tr cdc-owl:hasTrackingPermission 1 .
  ?tr cdc-owl:tracksLocationEntity ?trackLoc .
  ?trackLoc cdc-owl:hasTrackingPermission 1 .
  ?trackLoc cdc-owl:activeTracking 1 .
  ?trackLoc cdc-owl:hasTimeConstraint ?time .
  ?trackLoc cdc-owl:trackingNotificationOption ?option .
  ?trackLoc cdc-owl:isInside 0 .
  ?time cdc-owl:appliesTo ?timeInterval .
  ?timeInterval cdc-owl:intervalFrom ?i1 .
  ?timeInterval cdc-owl:intervalTo ?i2 .
  ?trackLoc cdc-owl:refersTo ?freqLoc .
  ?freqLoc a cdc-owl:FrequentLocationEntity .
  ?freqLoc cdc-owl:refersTo ?loc .

  FILTER (?i1 < "10:00:00"^^xsd:time) .
  FILTER (?i2 > "10:00:00"^^xsd:time) .
  FILTER (?option = 1 || ?option = 3) .
  FILTER (?loc IN (cdc:loc183))
}
```

Listing 8.6 Querying *contextdatacloud.org:8891/sparql-auth* for users tracking the frequently visited locations of `cdc:user00001` with enabled tracking permissions and constraints

If the query in Listing 8.6 delivers a set of positive results, we switch to the *WiFi Approach* of the *Semantic Tracking*. With the request in Listing 8.7, we retrieve all WiFi access points (`?wifi`) within the coverage area of `omn:cell2109543_5275_262_1` that cover the location entities of interest for `cdc:user00001`.

```
PREFIX cdc: <http://www.contextdatacloud.org/resource/>
SELECT ?wifi
{
  omn:cell2109543_5275_262_1 omn-owl:covers ?wifi .
  ?wifi a omn-owl:WiFiAP .
  ?wifi omn-owl:coversLocationEntity ?loc .

  FILTER (?loc IN (cdc:loc183))
}
```

Listing 8.7 Querying *sparql.openmobilenetwork.org* for WiFi APs that cover the location entities of interest for `cdc:user00001`

The result of this query is a set of WiFi access points, which are now used for matching the APs of the client.

8.2.2 Popular Places Finder

The *Semantic Tracking* process for self-referencing LBSs is highlighted in this section via the *Popular Places Finder* service of the *CDCApp* that notifies users about popular locations in their vicinity matching to a certain context. This service exploits interlinks to location entities of LCD.

For this example, we assume that the user is connected to the cell `omn:cell2159101_5275_262_1`. After retrieving all location entities for this cell from the OMN, the result set is used to query the LCD endpoint (see Listing 8.8) for matching these specific location entities to preferences and contextual constraints of the user. Only those location entities are selected having the category `cdc:restaurant` or `cdc:fastFood` with more than 10 check-ins occurred in the afternoon (between 12 p.m. and 6 p.m.) during weekdays when the temperature was between 10 and 25° C. The result is `cdc:loc473` with 16 check-ins that meets these parameters.

```
PREFIX cdc-owl: <http://www.contextdatacloud.org/ontology/>
PREFIX cdc: <http://www.contextdatacloud.org/resource/>
PREFIX gr: <http://purl.org/goodrelations/v1#>
SELECT ?poi, COUNT(*) AS ?checkInCount
{
  ?checkin a cdc-owl:CheckIn .
  ?checkin cdc-owl:refersTo ?poi .
  ?checkin cdc-owl:hasWeatherSituation ?weather .
  ?weather cdc-owl:hasTemperature ?temp .
  ?checkin cdc-owl:hasCheckInMinutes ?minutes .
  ?poi cdc-owl:hasLocationCategory ?category .
  ?checkin cdc-owl:occurredDuringDay ?day .

  FILTER (?minutes < 1080) .
  FILTER (?minutes > 720) .
  FILTER (?temp > 10) .
  FILTER (?temp < 25)  .
  FILTER (?category IN (cdc:restaurant, cdc:fastFood)) .
  FILTER (?day NOT IN (gr:Saturday, gr:Sunday)) .
  FILTER (?poi IN (cdc:loc11176, cdc:loc473, cdc:loc9590, cdc:loc3826,   cdc:
        loc8057, cdc:loc10668, cdc:loc453)) .
}
GROUP BY ?poi
HAVING (COUNT(*) > 10)
```

Listing 8.8 Querying *sparql.contextdatacloud.org* for location entities with contextual constraints

Assuming that the smartphone of the user measures the WiFi AP `omn:wifiap 1623563669` as the one with the highest signal (which is an indicator that the user is in the vicinity of this WiFi AP), he would be notified about the popular restaurant `cdc:loc473`.

8.3 OpenMobileNetwork Geocoder

Another demonstrator that we have implemented is the *OpenMobileNetwork Geocoder* Web interface[1] for directly comparing address query results of our *Semantic Geocoding* approach against *Nominatim* and *Google* (see Fig. 8.3).

The Web interface enables users to enter addresses (with or without attaching network topology data) located in Berlin, Helsinki, and San Francisco. The address examples that can be selected in the drop-down menu highlight various ambiguous address query scenarios. For Schillerstraße, for example, which exists several times in Berlin, *Nominatim* delivers wrong coordinates, whereas our *OMN Geocoder* and *Google* provide more accurate results.

[1]http://www.openmobilenetwork.org/geocoding/.

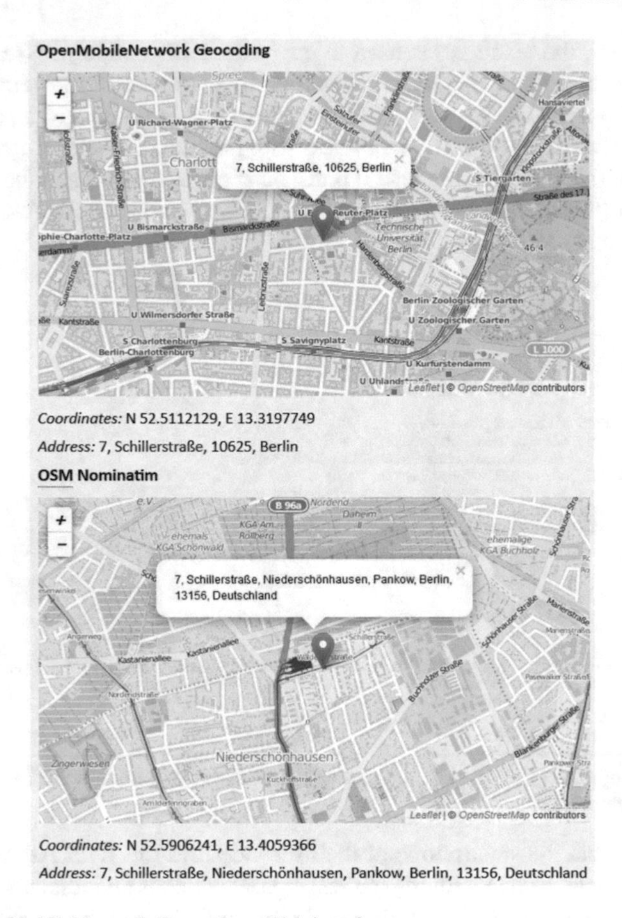

Fig. 8.3 OpenMobileNetwork Geocoder – Web interface

Furthermore, users of the Web interface are able to perform a live evaluation of our *Semantic Geocoding* in comparison to the competitors using the entire address datasets of the three exemplary cities. Here, each address can be tested in 4 different formats (with and without street number, city, or postal code). The results of the evaluation show the overall performance of each geocoding service with different types of address input.

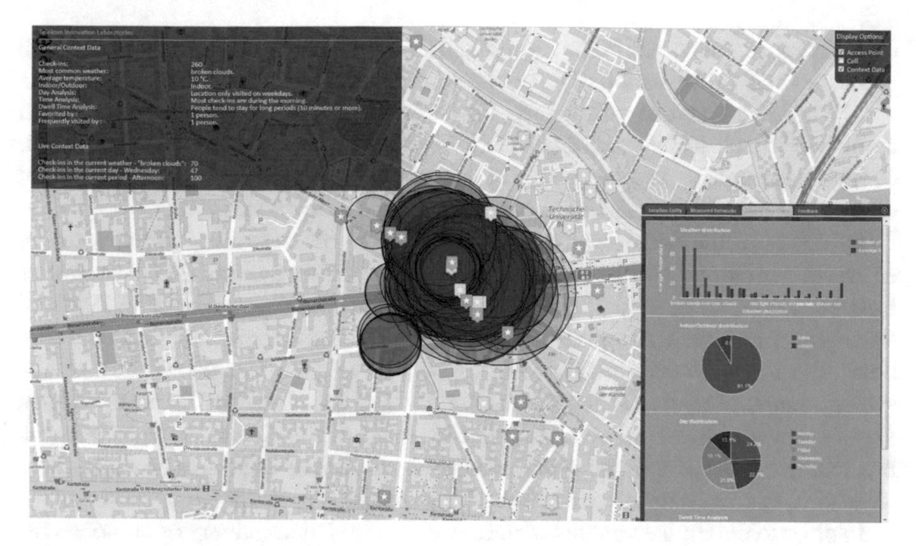

Fig. 8.4 Location Analytics Map

8.4 Location Analytics Map

The *Location Analytics Map*[2] (see Fig. 8.4) visualizes aggregated and rudimentally analyzed LCD data on a map using OSM. Each location entity is represented by a marker, while these markers are distinguished by different colors denoting the visiting frequency of a location entity. Red markers highlight places having more than 80 check-ins, whereas blue markers stand for locations with 40–80 check-ins. Green markers represent places in which people have checked in more than 10 times up until to 40, while grey is the color for location entities that were not visited at all.

Sliding over a marker pops-up the *General Context Data* window that shows a summary of the analyzed context aggregated through all check-ins. By checking the *Context Data* checkbox in the *Display Options*, this window is extended by *Live Context Data* information that matches the current context of the user to the analyzed context and counts how many past check-ins fit to his contextual situation. Furthermore, the checkboxes *Access Point* and *Cell* in the *Display Options* illustrate the coverage areas of the mobile network cells and WiFi APs (retrieved from the OMN) when sliding over a marker.

A click on a marker opens a window on the right side of the map, which shows detailed and aggregated information about this place by processing through all check-ins given for it. The *Location Entity* tab lists all static and domain-specific RDF information that is related to the location entity, whereas the *Measured Networks* tab highlights all mobile network cells and WiFi APs that were collected for this place during all check-ins. The *Context Data Charts* tab illustrates the distribution of the

[2]http://map.contextdatacloud.org.

diverse context data types in the form of charts, whereas the *Feedback* tab shows all comments and a distribution of the given ratings.

The collected context information for all check-ins is analyzed as follows: The number of check-ins for a location entity is obtained directly from the SPARQL endpoint through a simple query for the count of `cdc-owl:CheckIns` that `cdc-owl:refersTo` a given `cdc-owl:LocationEntity`.

Furthermore, the `cdc-owl:WeatherSituation` is retrieved (if provided) for each check-in using the `cdc-owl:hasWeatherDescription` and `cdc-owl:hasTemperature` properties. This information is grouped by the weather descriptions and the average temperature is calculated for each of these descriptions. The *Most common weather* corresponds to the weather description and its average temperature that was collected for most of the check-ins. A more detailed view is provided in the *Weather Distribution* chart of the *Context Data Charts* tab illustrating the list of weather descriptions, the number of check-ins in which these weather descriptions have been observed and the average temperature for them.

In order to determine whether a place was rather visited indoors or outdoors, the endpoint is queried for the `cdc-owl:isIndoor` flag given for all check-ins that refer to this location and the relation (in percentage) of how many check-ins to the total are considered indoor or outdoor is calculated. The total distribution is shown in the *Indoor/Outdoor Distribution* chart, whereas the one that has a bigger proportion is given as a summary in the *General Context Data* window.

The same calculation is applied for analyzing the day distribution of check-ins. Each check-in is associated with the `gr:DayOfWeek` when it occurred. The relation between the amount of check-ins for each day and the total of check-ins is shown in the *Day Distribution* chart, while an aggregated and human-readable description of the day distribution is displayed in the pop-up window.

The time analysis is performed by spliting a day into four different periods and associating these periods with textual descriptions. 12 a.m. to 6 a.m. corresponds to *night*, whereas *morning* is between 6 a.m. and 12 p.m. The timeframe between 12 p.m. and 6 p.m. matches the *afternoon* and *evening* is defined between 6 p.m. and 12 a.m. For each check-in, the value of the hour (determined via `cdc-owl:hasCheckInMinutes`) is mapped to one of the four periods and the textual description with the most check-ins is shown in the *General Context Data* window.

By calculating the difference between the values for `cdc-owl:hasCheckInMinutes` and `cdc-owl:hasCheckOutMinutes`, the dwell times of all check-ins for a location entity are determined. In a next step, these dwell times are separated into long visits (the ones that are longer than 10 min) and short visits (the ones that are shorter than 10 min). Furthermore, the average dwell time for all check-ins as well as individually for each of the two groups is calculated. In the *Dwell Time Analysis* chart of the *Context Data Charts* tab, the amount of check-ins for each of the three cases (total, long and short visits) and their corresponding average dwell times can be seen. The *Dwell Time Analysis* in the pop-up window, on the other hand, displays the group of visits (i.e., long and short visits) that has the biggest amount of check-ins.

Information about `cdc-owl:FavoriteLocationEntitys` and `cdc-owl:FrequentLocationEntitys` are retrieved from our private SPARQL endpoint and shown in the pop-up window in case a location entity is frequently visited or favorited by users.

In order to match the current context of a user to the historic context related to the check-ins, available and up-to-date context data is retrieved as soon as the user clicks on the *Context Data* checkbox in the *Display Options*. First, *OpenWeatherMap* is queried for the current weather condition. The textual description of this weather condition is matched to the amount of check-ins that refer to the location entity having this specific weather description. In addition, the timestamp of the CDC server is used to determine the current day of the week and time of the day. This information is also matched with the check-ins for the place that occurred on that day and fit to one of the time periods defined above. The results are shown in the *Live Context Data* part of the pop-up window.

Part III
Evaluation

Chapter 9
Crowdsourced Network Data Estimation Quality

Services that are based on the geographic mapping model of the *OpenMobileNetwork* strongly depend on the accuracy of the estimated network topologies including the positions and coverage areas of the mobile network cells and WiFi access points. Mobile network operators obviously possess the most accurate network topology data since they are responsible for the deployment of their infrastructures.

Within this evaluation, we defined the real topology data of a mobile network operator as the gold standard reference and performed an estimation quality analysis by comparing the geodetic distance of the mobile network cells within the *OpenMobileNetwork* to the real positions of these cells. We performed this evaluation twice within a time frame of approximately 1.5 years in order to see if the estimations led to better results with more collected network measurements over time.

9.1 Crowdsourcing Statistics

As of March 20th, 2015, the *OpenMobileNetwork* dataset comprised crowdsourced network data covering 76 countries all over the world. Table 9.1 gives an overview about the collected mobile network data.

So far, 1,253,705 measurements have been acquired for Cell-IDs in total out of which 478,687 were collected with *OMNApp* as well as *Jewel Chaser* and 775,018 using the *CDCApp*. Most of these measurements have been collected in Germany, France, Monaco, the United States, Turkey, Finland, and Saudi Arabia, with all of these countries having more than 30,000 measurements respectively. During data processing, 21,190 unique cells were calculated consisting of 5,900 2G and 15,290 3G cells. In Germany, 4,816 cells are operated by *Telekom*, whereas *Vodafone* deployed 1,374 cells. 2,782 cells belong to *E-Plus*, while O_2 has 2,316 cells.

A. Uzun, *Semantic Modeling and Enrichment of Mobile and WiFi Network Data*, T-Labs Series in Telecommunication Services,
https://doi.org/10.1007/978-3-319-90769-7_9

Table 9.1 Statistics of collected mobile network data via crowdsourcing

Continent	Countries	Network Providers	Measurements	2G Cells	3G Cells
North America	9	17	127,689	696	1,910
South America	6	16	126	9	30
Asia	25	75	185,930	1,074	3,842
Europe	25	67	933,660	4,097	9,317
Africa	10	13	3,297	23	138
Oceania	1	3	3,003	1	53
Sum	76	191	1,253,705	5,900	15,290

Table 9.2 Statistics of collected WiFi network data via crowdsourcing

Continent	Measurements	WiFi APs
North America	1,375,463	76,764
South America	251	62
Asia	850,896	48,687
Europe	9,229,805	270,283
Africa	6,393	797
Oceania	9,121	421
Sum	11,471,929	397,014

Table 9.2 highlights the number of collected WiFi measurements worldwide. In total, we have collected 11,471,929 WiFi measurements out of which we have determined 397,014 unique WiFi access points. *OMNApp* and *Jewel Chaser* provided 2,962,347 measurements, whereas 8,509,582 were conducted using the *CDCApp*.

The huge differences in the number of collected measurements between the apps are due to the design of the data collection function. *OMNApp*, for instance, uses a warwalking and wardriving method to collect network data. Users explicitly click on a button to start and stop the measure mode whenever they go to places where no measurements were made before, for example. *Jewel Chaser*, on the other hand, utilizes gamification methods in order to provide incentives to users, so that they contribute with their data. Nevertheless, this app also sends network data only whenever the users actively play the game leading to a limited number of measurements throughout the day. Both of these apps are rather designed to systematically deliver measurements with high accuracy based on GPS.

The data collection function of the *CDCApp*, however, runs non-stop every three minutes as soon as users install the app generating a huge number of measurements over time and quickly extending the coverage of the *OpenMobileNetwork*. But this approach entails some disadvantages: First, we have to rely on the less accurate *Network Location Provider* of *Android* instead of GPS, because an exhaustive usage of GPS will lead to high smartphone battery consumption. Another aspect is that GPS will not work constantly since users do not always reside in an outdoor environment.

In addition, a lot of the measurements will be collected at the same spot (while users are sleeping, for example) having no significance at all for the network topology estimation.

Therefore, in order find a good balance between high accuracy estimations and more coverage in the *OpenMobileNetwork*, we separated the measurements in the database based on their origin. If we have collected measurements for a specific cell or WiFi access point with the *CDCApp* that we have not acquired yet by using the other apps, we use the measurements for estimation in order to extend the coverage of the *OpenMobileNetwork.* This approximated cell or WiFi access point has an `omn-owl:providedByCDC` predicate attached to it. However, as soon as *OMNApp* or *Jewel Chaser* provides measurements for the same cell or WiFi access point, we remove all *CDCApp* measurements from the estimation process and recalculate the position of this cell or WiFi access point by only incorporating the accuracte GPS measurements.

9.2 Accuracy of the Position Estimation

We performed two separate evaluations within a time frame of approximately 1.5 years and used different datasets of real mobile network data for these analyses provided by an operator. The first evaluation comprised 64 random mobile network cells with GSM access technology, while the recent comparison was done on a dataset consisting of all 2G and 3G cells in Berlin, Germany. Taking the dataset of the operator as a reference, which included a list of Cell-IDs, LACs, and WGS84 coordinates, we filtered our *OpenMobileNetwork* dataset accordingly to only comprise the same cells as the gold standard.

9.2.1 Distance Comparison for 64 Random Mobile Network Cells

For the first evaluation performed in 2013 [C5, J1], we used all four position estimation algorithms analyzed in Sect. 4.4 and applied them on the evaluation dataset of 64 mobile network cells in order to calculate the geodetic distances between the estimated positions and the reference positions. Figures 9.1 and 9.2 highlight the results for all four approaches. The x-axis of the charts list all 64 GSM cells, whereas the y-axis illustrate the distances to the real position of the cells given in [km]. The cells on the x-axis are sorted ascending according to the distances of the standard centroid-based approach.

Figure 9.1 displays the evaluation results for the standard and weighted centroid-based approaches. The results for the standard centroid-based approach show that for the first 31 cells (being half of the evaluation dataset), the distances are shorter than

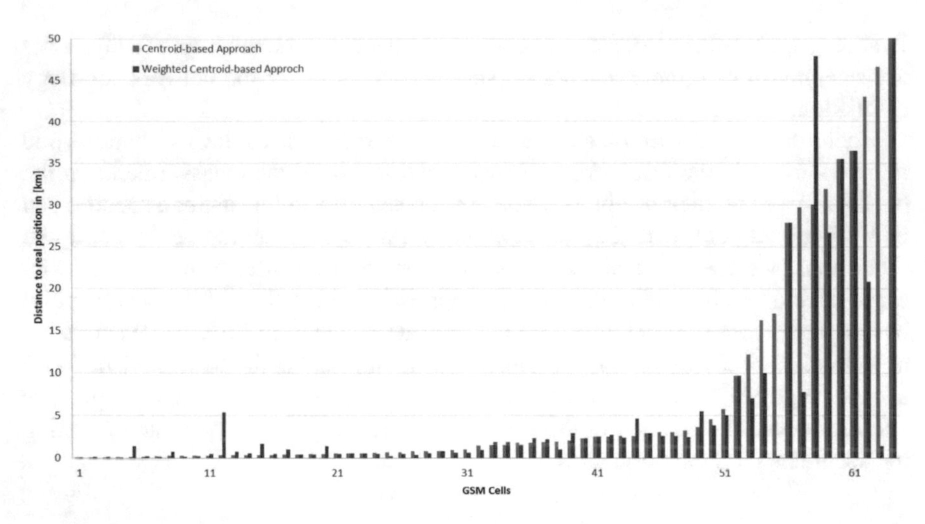

Fig. 9.1 Standard versus Weighted Centroid-based Approach

1 km. This indicates that we can already achieve acceptable results by only relying on a simple algorithm. The next 19 cells (between cell #32 and #50) are up to 5 km away from their real positions, while the last 12 cells (#53 to #64) have much longer distances (up to 50 km).

Using the weighted centroid-based approach, we have achieved a significant decrease for 7 of the last 12 cells proving the positive impact of the weighted signal strength parameter. Overall, the results are mostly comparable to the standard centroid-based approach, but with shorter and fluctuating distances for most of the cells.

Figure 9.2 comprises the results of the minimum enclosing circle as well as the grid-based approach. For the minimum enclosing circle approach, we can see that the results are very similar to the standard centroid-based approach. This behavior is due to the similar distribution of measurements for most of the cells within the *OpenMobileNetwork*. The measurements are located in a linear shape along a street, for example. Therefore, the problem of inhomogeneously distributed points, which we described in Sect. 4.4, does not occur. As a result, both approaches perform very similar in this evaluation.

Applying the grid-based approach, on the other hand, delivers the best results in terms of distance. The distances of 33 cells are below 900 m, whereas 22 cells have estimated positions that are between 1 and 5 km away from the real cell positions. Especially, the last 20% of the distances were significantly shortened. We have only four cells that have a distance more than 10 km.

Table 9.3 summarizes the results of all four approaches in terms of average and median distance as well as average distance after discarding 10% of the results with the longest and the shortest distance.

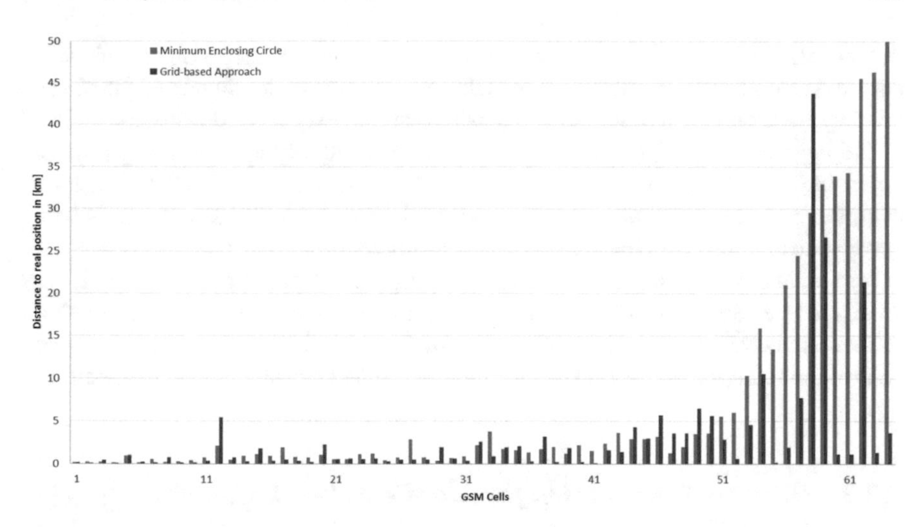

Fig. 9.2 Minimum Enclosing Circle versus Grid-based Approach

Table 9.3 Overview of the evaluation results

	Arithm. mean	Median	Trim. mean
Standard centroid-based approach	7,325.39 m	1,499.01 m	4,005.82 m
Weighted centroid-based approach	5,829.08 m	1,409.87 m	3,538.38 m
Minimum enclosing circle	7,190.41 m	1,568.37 m	3,717.92 m
Grid-based approach	3,094.82 m	801.48 m	2,710.41 m

Clearly, the grid-based approach provides the most accurate results followed by the weighted centroid-based approach, which performs quite similar. However, analyzing the results of all algorithms, we recognized a pattern of increasing distances for the last 20% of the dataset. Having a deeper look at the crowdsourced dataset of the *OpenMobileNetwork* and comparing it with the real operator data, we realized that base stations equipped with repeaters result in these unusually large distances. Multiple repeaters are deployed along highways or subways, for example, in order to extend the coverage area of a base station without changing the Cell-ID in the extended area. Hence, smartphones are always connected to the same Cell-ID when being in the vicinity of these repeaters (e.g., when driving on a highway from one city to the other) and conducting network measurements, so that the whole coverage area consisting of this base station and multiple repeaters is treated as one huge cell by the position estimation algorithms of the *OpenMobileNetwork*. Centroid-based approaches perform poorly in these scenarios, because the position is estimated as the (weighted) centroid based on all network measurements distributed over this huge distance. The effect of this problem on the grid-based approach, however, is weaker, because it uses the highest concentration of measurements with a strong signal strength for the estimation (and approximates either the base station or one of the repeaters as the position). Another aspect is that network topology is subject to constant change. New base stations for UMTS and LTE are deployed and GSM cells are degraded. Outdated data may also easily result in very large deviations.

Taking the achieved results as a reference, we initially applied the grid-based algorithm as our position estimation approach for the *OpenMobileNetwork*. However, the grid visualization on the map with all cells and WiFi access points being systematically positioned side-by-side and additionally several cells as well as WiFi access points overlapping each other on the exact same position did not look very "realistic" (see Sect. 4.4.6). Therefore, in the course of time, we switched the position estimation as well as the map visualization to the second-best performing weighted centroid-based approach and used this approach solely for our second evaluation.

9.2.2 Distance Comparison for Mobile Network Cells in Berlin

In 2015, we ran our second evaluation utilizing a much bigger dataset of the operator with network topology data for the whole city of Berlin, Germany. Comparing the operator's dataset to the crowdsourced data of the *OpenMobileNetwork*, we were able to identify and filter 355 2G and 519 3G cells for Berlin to be used within the evaluation. Again, we calculated the geodetic distances of our estimated cell positions to the real positions given by the operator in order to analyze whether more network measurements increase the accuracy of the (weighted centroid) position approximations over time.

Figure 9.3 shows a plot of the distance calculation results in the form of a histogram. Here, the x-axis of the histogram represents distance thresholds given in [m], whereas the y-axis lists the number of cells that are within these distance intervals. Having a deeper look at the blue bars, we can see that more than half of the 355 2G cells (being 180 exactly) are only up to 500 m away to their real positions, whereas 70 cells have a distance below 1 km. 29 other cells have estimated positions with max. 1.5 km deviation, while 70 cells range from 2 to 5 km. A small number of 6 cells is more than 5 km away.

The red bars, on the other hand, represent the calculated distances for the 519 3G cells in Berlin, which are far better approximated than the 2G cells. Most of the cells (being 395) have an estimated location below 500 m, whereas 75 cells are up to 1 km and 16 cells are up to 1.5 km away to their real positions. 16 other cells range between 2 and 5 km, while only one has a very long distance (up to 10 km).

In contrast to our first evaluation in Sect. 9.2.1, the results in our second analysis clearly illustrate that most of the cells have been estimated with a good accuracy. This indicates that more network measurements have a positive impact on the estimations. We were able to decrease most of the large distances of up to 50 km that were calculated in the first run. The worst distances that we have computed in the second evaluation are below 10 km. We strongly believe that there is even more potential to

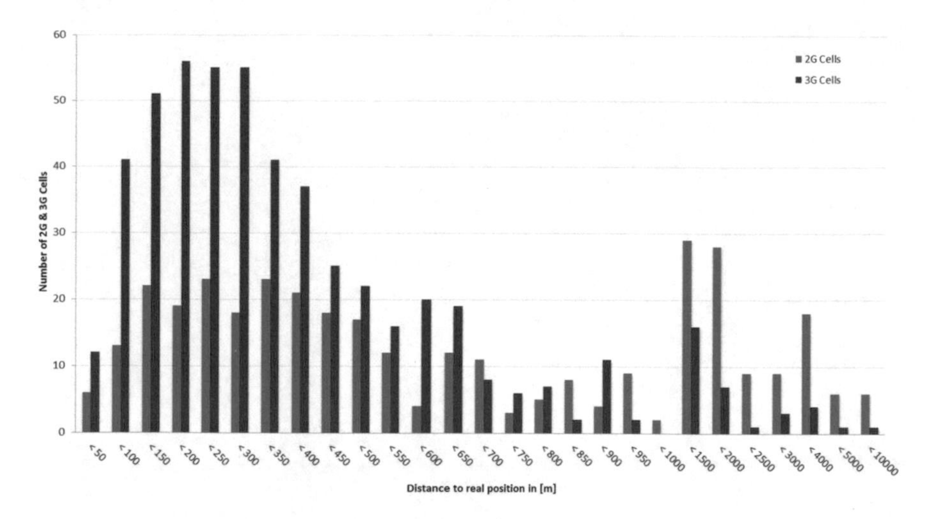

Fig. 9.3 Distance Histogram for 2G and 3G Cells

improve the results if we had more network measurements distributed all over the city of Berlin. At the moment, we have parts in Berlin, which have a better coverage by network measurements due to frequent visits by users of the *OpenMobileNetwork* measurement clients (e.g., our office area) than other spots in the city. The distribution of the measurements as well as the problem with the repeaters mentioned in Sect. 9.2.1 lead to the calculation of the large distances of up to 10 km.

All in all, the evaluation of the network data estimation quality proves that context-aware services based on the geographic mapping model of the *OpenMobileNetwork* rely on a solid and accurate network topology dataset.

Chapter 10
Applicability of Services

This section highlights the applicability of the services that are based on semantically enriched mobile and WiFi network data. For this purpose, we have performed in-depth accuracy evaluations for the *Semantic Tracking* as well as the *Semantic Geocoding* respectively that were introduced in Sects. 7.2.1 and 7.2.2. The accuracy of the *Semantic Tracking* function is evaluated in terms of a comparison of the distance between the location when a friend tracking notification upon entering (or leaving) a favorite location entity is received and the actual position of the POI, whereas the performance of the *Semantic Geocoding* is compared against its competitors in terms of ambiguous or incomplete input data.

10.1 Semantic Tracking: Distance Calculation

In LBSs, the positioning accuracy is a crucial factor and is of extraordinary importance. Even though the requirements of accuracy differ from one LBS application to the other, a certain level of accuracy is definitely required to ensure correct functionality and an acceptable LBS experience. For this purpose, our *CDCApp* has been used to evaluate the accuracy of the *Semantic Tracking* function through conducting field trials. The main metric being analyzed is the distance between the tracked user's current position at the moment the tracking party gets notified by an *entering* or *leaving* event and the position of the location entity itself within LCD.

Three test users with *Android* devices have used a specifically developed evaluation version of the *CDCApp* by creating a profile, adding each other according to a friend relation, subscribing for the service and issuing tracking requests to each other. They have defined 51 favorite location entities in total to be tracked by each other in four different districts of Berlin, Germany (i.e., Wedding, Charlottenburg,

A. Uzun, *Semantic Modeling and Enrichment of Mobile and WiFi Network Data*, T-Labs Series in Telecommunication Services,
https://doi.org/10.1007/978-3-319-90769-7_10

Schöneberg, and Steglitz), where there is a significant concentration of location entities and good network coverage.

In the evaluation version of the app, which is slightly different than the official release, GPS is always turned on for accuracy reasons and whenever the server detects an *entering* or *leaving* event of the tracked user, it issues a notification requesting his accurate GPS position and a number of additional parameters. This information is logged into a CSV file on the server and used for calculating the distance between the tracked user's current position at the moment the tracking party receives an *entering* or *leaving* notification and the position of the favorite location entity in LCD. The collected parameters include:

- A unix epoch timestamp
- The name of the service being evaluated
- The name of the location entity that is subject to notification
- The ID of the location entity
- Event status: `1` for an *entering* event, `2` for a *leaving* event
- WiFi or Cell-ID approach
- Latitude value for the position of the user determined via GPS
- Longitude value for the position of the user determined via GPS
- The calculated distance
- The ground area of the location entity (given as a radius value in meters in combination with the position of the location entity)
- The calculated distance decreased by the radius value of the ground area
- The accuracy of the GPS position
- The number of APs that cover the location entity
- The number of APs considered for the tracking service

In order to ensure high quality and accuracy in the evaluation data when calculating user positions via GPS, we reserved time slots in which we made sure that the test users are walking outdoors and that no other test client is running. Each user performed five walking tours on five days for the district(s) assigned to him, which lasted between one and two hours. For each tour, we have changed the number of WiFi APs taken into consideration when performing the *WiFi Approach* of the *Semantic Tracking* (see Sect. 7.2.1).

On the first tour, we only considered the one WiFi AP with the strongest signal strength that covers the favorite location entities to be tracked, whereas the second and third tour respectively included the two and three APs with the highest signal strengths where we checked whether the location entities to be tracked were covered by at least one of them. The forth and fifth tour also included two and three APs. Here, however, we sent only notifications to the tracking party whenever the favorite location entities were covered by all (two or three) APs at the same time.

For simplicity reasons, the position of a location entity in LCD is represented by WGS84 coordinates as the central point of the building and a radius value in meters defining the ground area of this location entity as a circle. For the distance calculation between the tracked user and the location entity of interest, we first took the central

WGS84 position of the location entity. However, when analyzing the evaluation data after our tours, we realized that especially buildings with a large ground area led to significantly false distance calculations: Even though the user was right next to the building (e.g., across the street) when the tracking party received the notification, the distance (from across the street) to the center point of the building negatively affected the overall distance calculation. Since we were not interested in the distance from the user to the center of the location entity, but rather to the closest point of the area that defines the location entity, we decreased the value of the ground area radius from the calculated distance in order to obtain the shortest distance from the user position to the location entity representation. Sometimes, however, the tracked user was already inside the ground area circle that also covered parts of the street in front of the building, for example, causing negative values when substracting the radius from the user's distance. In such cases, we took 0 m as a defined value for the result of the distance calculation.

Furthermore, we also took the accuracy value of the GPS position into consideration that defines a radius of a circumference that the user is supposed to be inside. We discarded all measurements where the GPS position had an accuracy value higher than 30 m assuming that such inaccurate values cannot hold as a reference for our calculations.

Figure 10.1 illustrates the results for the user distances to the location entities that are covered by the AP with the highest measured signal strength, whereas

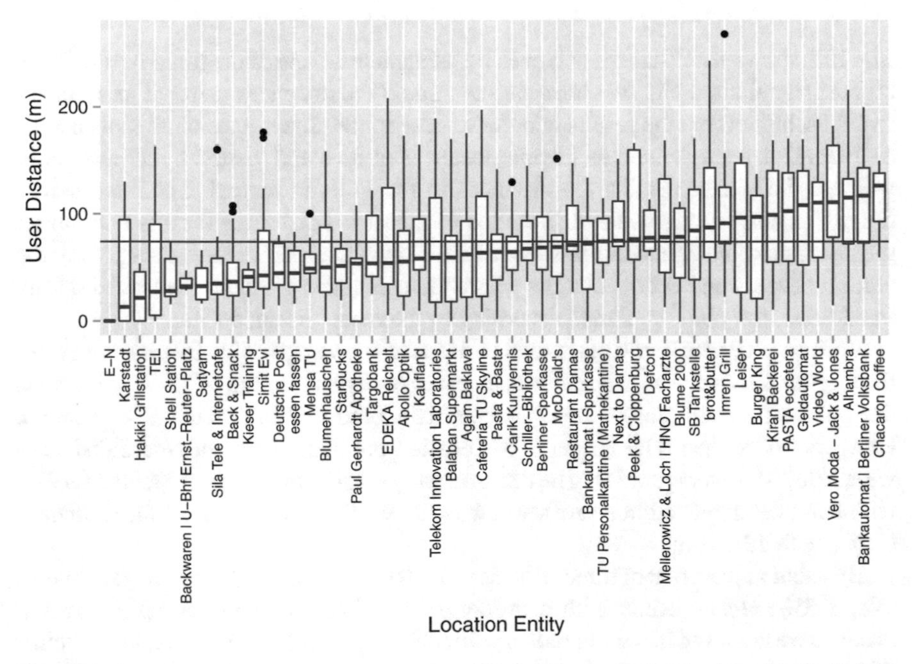

Fig. 10.1 Semantic Tracking – user distance to POI that is covered by the AP with the highest signal strength

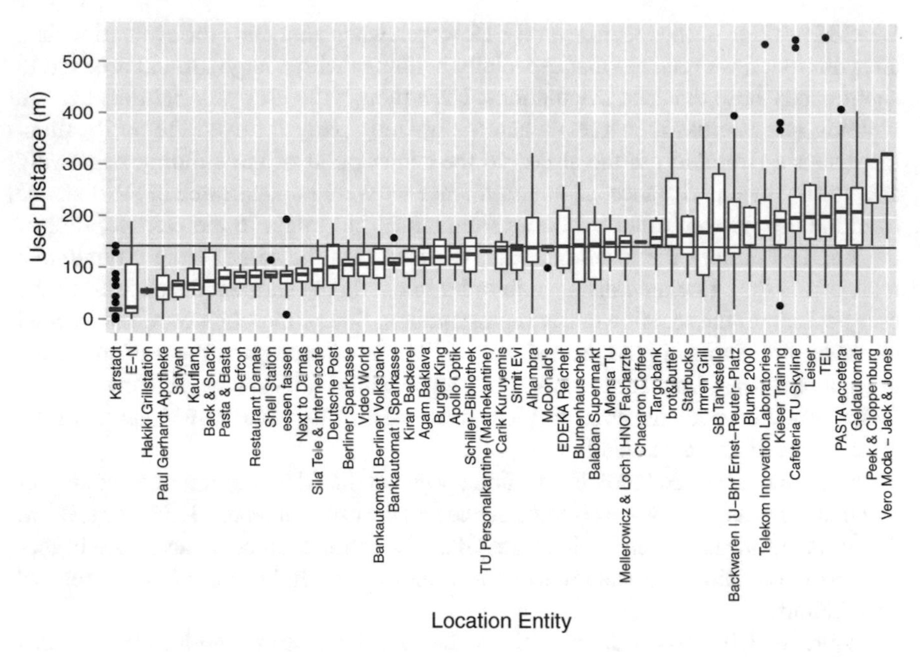

Fig. 10.2 Semantic Tracking – user distance to POI that is covered by at least 1 of 3 APs with the highest signal strength

Fig. 10.2 shows the distances achieved by using at least one of the three APs with the highest signal strengths. We have also evaluated the distances using at least one of two APs, but left out the results in order to save space. The x-axes of all plots list the 51 POIs that were defined as favorite location entities and visited by all test users, whereas the y-axes display the distances to the POIs in meters. The interquartile range of a box represents the distribution of measured distances for this POI during the test run excluding the outliers. For a better visualization, we have only plotted the *entering* events and cut off outliers where the calculated distances exceeded 600 m. In addition, the plots are sorted according to the mean distance value.

With one AP (see Fig. 10.1), a total of 518 and an average of 10.15 *entering* events throughout the first tour, we achieved an average distance value of 73.98 m between the point where the notification has been triggered and the actual POI location. However, more than 60% (equal to 32 location entities) of the measured distances were below the average value. The best accuracy result was achieved for the POI *E-N* with 0 m (see explanation above), whereas the worst result occurred for *Chacaron Coffee* with 126.10 m.

By using at least one of three APs (see Fig. 10.2) on our third tour, we generated a total of 558 *entering* events with an average of 10.94 notifications per POI. The mean distance we achieved here is 141.29 m with 53% (equal to 27 of 51 POIs) being below this value. We measured the best user distance for *Karstadt* with 19.39 m, whereas *Vero Moda - Jack & Jones* delivered the worst result with 323.34 m.

The results in Figs. 10.1 and 10.2 clearly show that there is an inverse relation between the number of considered APs and the accuracy of the *Semantic Tracking* approach: The more APs we use in a unified manner for detecting location entities of interest, the bigger is the distance between the user and the POI. This is due to the fact that if we consider n APs with the strongest RSSI, a position calculation results in p location entities covered by the n APs leading to the conclusion that the client is "inside" the p location entities. If we consider $n + 1$ APs with the strongest RSSI, the position calculation results in $p + q$ location entities where q represents the location entities covered by the additional AP. As this additional AP has a weaker RSSI than the other n APs, this AP is supposedly located further away and covers therefore location entities that are further afar. This leads to the fact that the user is considered to have entered further location entities with $n + 1$ APs in comparison to n APs.

Furthermore, the interquartile range of the boxes and hence the distribution of the distance calculations strongly depends on the quality of the given WiFi network coverage data in the *OpenMobileNetwork*. During our evaluation, we realized that some of the WiFi network data within the OMN had links to POIs that are normally not within the coverage area of this AP (e.g., `omn-owl:wifiap-544168409`). This could have many reasons: First, the data is collected via crowdsourcing, which is always error-prone to some degree. Another reason could be that the AP has dynamically "changed" its position due to new crowdsourced measurements, but the

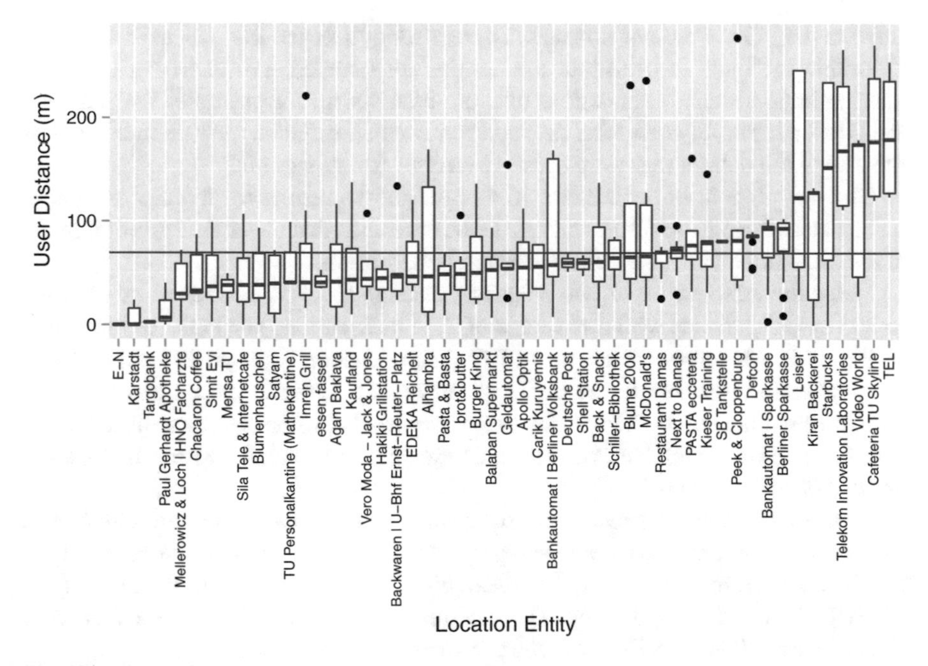

Fig. 10.3 Semantic Tracking – user distance to POI that is covered by 2 APs at the same time

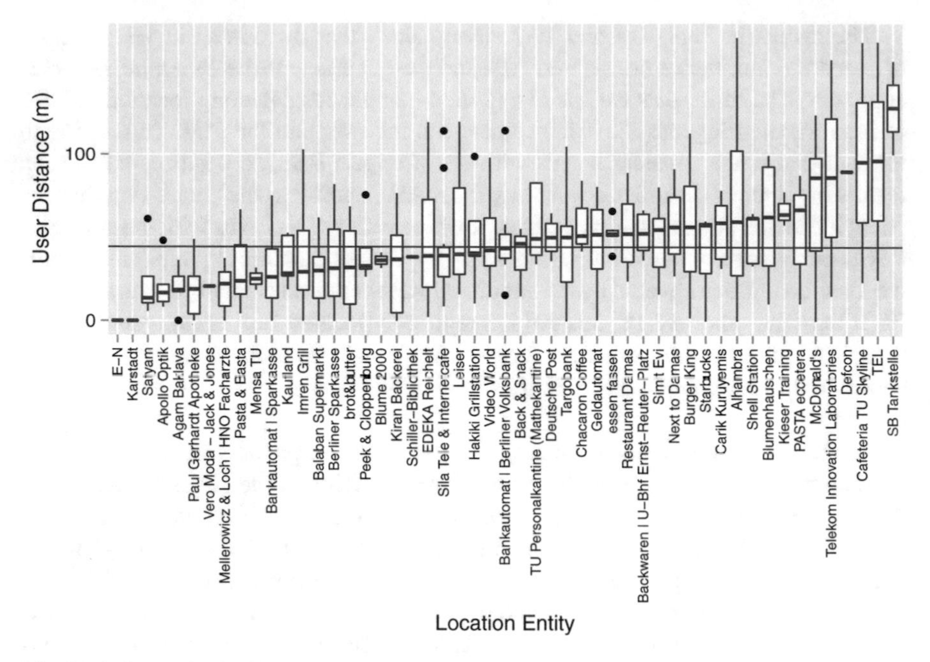

Fig. 10.4 Semantic Tracking – user distance to POI that is covered by 3 APs at the same time

links to the POIs were not updated. These wrong links led the *Semantic Tracking* algorithm to "find" favorite location entities for scanned APs at wrong locations causing large distance calculations of more than 400 m, for example. We analyzed the OMN dataset for such false data and removed all distance measurements out of the evaluation dataset that were caused by APs with wrong links.

The box plots in Figs. 10.3 and 10.4 highlight the results for the user distances using two or three APs with the highest signal strengths that cover the POIs of interest at the same time. Here, we achieved much better results.

Using the intersection of two APs for detecting location entities of interest led to 395 *entering* events with an average of 7.74 notifications per POI. As illustrated in Fig. 10.3, we scored an average of 69.25 m distance between the point where the notification was sent to the tracking party and the tracked user's position with almost 70% (equal to 35 POIs) being below the average value. The best distance was calculated for *E-N* with 0 m, whereas the worst result was achieved for the location entity *TEL* at 179.07 m.

Figure 10.4 shows the results for the *Semantic Tracking* evaluation with 3 APs used at the same time. Here, we have generated 329 events with an average of 6.45 notifications per location entity. The mean distance value is 44.13 m with 49% (i.e., 25 POIs) below the mean value. The best result was again achieved for *E-N* with 0 m, whereas *SB-Tankstelle* generated the worst result with 128.05 m.

Figures 10.3 and 10.4 clearly indicate that the more APs we intersect for detecting POIs of interest, the better the results become. These findings are very logical since

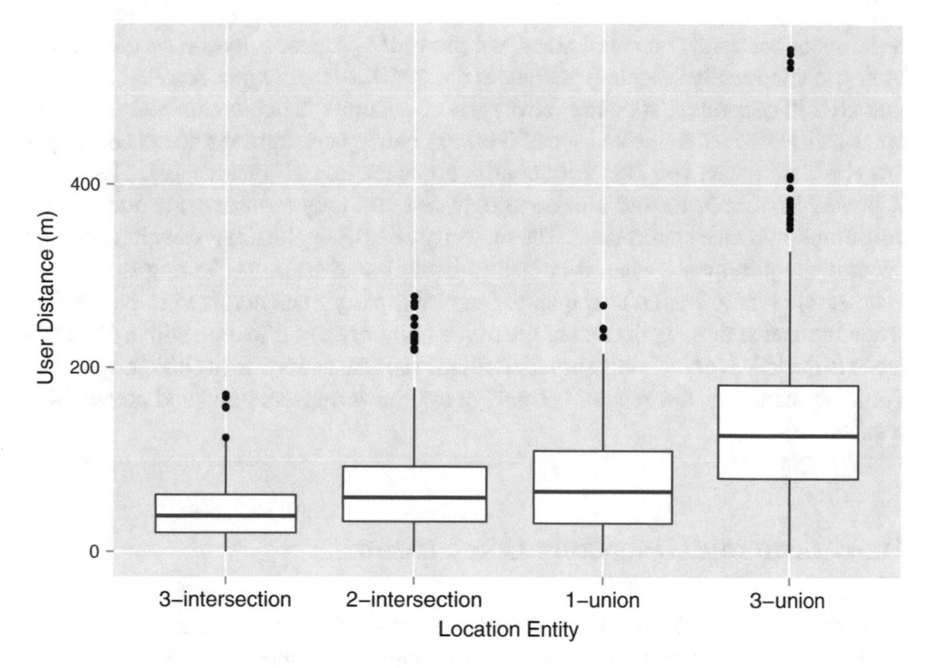

Fig. 10.5 Semantic Tracking – overall comparison

we ensure that events are only generated whenever a sophisticated fingerprint of WiFi APs in combination with a list of POIs is given. Theoretically, we can conclude that the more sophisticated this fingerprint is, the better is the distance accuracy. However, we need to make sure that we find a good balance between a sophisticated fingerprint and the desired (as well as required) accuracy. Because, the more sophisticated a fingerprint is, the more it is difficult for phones to determine this fingerprint and hence send (entering) event notification. The dropping number of created events per tour is an indication for this claim.

Figure 10.5 is an overall comparison of the user distances calculated on four of our five tours. This plot verifies that for the *Semantic Tracking* approach, the number of considered APs and their usage is crucial for ensuring positioning accuracy in self-referencing as well as cross-referencing LBSs. Using a WiFi scan of a smartphone in order to identify POIs that are covered by at least one of the APs in the list (see *3-union* with 141.29 m and *1-union* with 73.98 m) delivers worse results than intersecting a certain number of APs that cover the location entities of interest at the same time (see *2-intersection* with 69.25 m and *3-intersection* with 44.13 m).

The evaluation clearly illustrates that semantically enriched mobile network and WiFi access point topology data in combination with links to other (geospatial) datasets can be a good alternative to classic geofencing methods in areas with dense network coverage (e.g., cities). In rural areas, the coverage areas of cells are much bigger and WiFi APs are not distributed as much as in the city. Here, the *Semantic Tracking* approach will not be able to completely replace geofencing methods in terms of accuracy. For rural areas, we therefore propose a hybrid solution combining (semantically modeled) geofences and network topologies. Another aspect is that GPS in general is, of course, always more accurate than our *Semantic Tracking* approach. However, the sole usage of it is very battery-consuming and hence a major drawback of proactive LBSs with continuous background tracking [9]. Therefore, it is very important to find a balance between accuracy requirements and battery consumption on the smartphone. The majority of LBS applications only have coarse accuracy requirements when it comes to background tracking. Taking the *Friend Tracker* application as an example, the tracking party does not need to be notified when the user is directly in front of the office building, a notification with a 45 m distance to the POI is completely sufficient for most of the users from a QoE perspective. For such scenarios, the *Semantic Tracking* approach delivers good and competitive results.

10.2 Semantic Geocoding: Comparison

The performance of a geocoding service strongly depends on the correctness and completeness of the input address data. Incomplete or ambiguous data, such as a street name without the postal code that exists several times in a city (e.g., Berliner Straße in Berlin, Germany), could often lead to wrong results in the geocoding process. Another example are queries that only contain a city name with multiple equally valid answers.

Therefore, we performed two evaluations using different benchmark address sets for Berlin, Germany, as the city provides very good mobile network coverage within the *OpenMobileNetwork*. The address sets have been created using *Bing*[1] in order to gain independence from the services to test against.

At first, we compared our *Semantic Geocoding* approach against *OSM Nominatim* and *Google* using a dataset of "conventional" addresses created randomly. This dataset included a majority of addresses that are uniquely available within the city. For the bounding box of Berlin specified with `(52.34163213095788, 13.084716796874998)` as the south-west and `(52.660142414421955, 13.77754211425781)` as the north-east corner, 5000 WGS84 coordinates have been randomly generated. These coordinates were then reverse geocoded with *Bing* leading to various types of addresses. Due to the fact that we only wanted to test fully specified addresses, we further post-filtered the results, so that each address contained a street name, a house number as well as the city name of Berlin excluding highways and squares with no house numbers. At the end, our dataset included 1186 distinct addresses.

The second analysis focused on "special address cases", i.e., ambiguous addresses existing several times in the city. For this purpose, the second dataset contained only street names that occurred at least twice in Berlin with different postal codes. These

[1] https://msdn.microsoft.com/en-us/library/ff701713.aspx

addresses were obtained from LGD as their SPARQL endpoint easily enables such requests. The street names have been geocoded with *Bing* in order to obtain WGS84 coordinates to compare to. For some of them, *Bing* returned the same WGS84 coordinates (as the center point of a street) even though they were queried with different postal codes. This indicated that the street was exactly the same located in a different postal code region being a candidate to be filtered out. In a next step, the remaining streets were filtered again by removing those that only occurred once with no ambiguity. In order to obtain fully specified addresses for the extracted coordinates, again a *Bing* reverse geocoding was performed for retrieving a house number. For those streets where this geocoding request failed, we included the nearest house number manually by checking *Bing* maps. This dataset of special cases comprised 476 distinct addresses.

Since our *Semantic Geocoding* focuses on the mobile geocoding use case, the evaluation strategy simulated a user searching for a target address (being one within the created datasets) from a location in the vicinity. For this purpose, we searched for (parts of) each address within both datasets from a random point up to 5 km away from this address. We used the coordinates of this random position in order to determine the list of mobile network cells from the OMN that cover this point assuming that a user would be connected to one of those mobile network cells in reality. Here, we decided to use the one cell for covering the random point whose range is closest to the average range of 1983 m determined from all cell ranges in the OMN. This cell could, but does not necessarily cover the target address.

A geocoding is then performed by first checking whether the target address is among those that are covered by the mobile network cell the user is theoretically connected to. If it is not covered by this cell, the target is crawled among those addresses that are covered by cells sharing the same LA as the current cell of the user. If the fully specified address was not found, perhaps due to missing house number information, the address parts were gradually removed until only the street name was searched.

In order to provide a fair comparison, we used the above defined bounding box for Berlin also for the *Google* and *OSM Nominatim* geocoding services. For each dataset, we ran three separate passes providing each geocoder with

- the full address (street name, house number, postal code, and city),
- partial address (street name and house number),
- street-level address (only street name).

Figure 10.6 illustrates the results for the distances calculated between the coordinates of the geocoding responses determined for each geocoding service and the coordinates of the target addresses within the randomly created dataset comprising a majority of uniquely available addresses. In contrast to *Google* and *Nominatim* where we have only used the bounding box of Berlin, the geocoding request of the *OMN Geocoder* was enriched with the mobile network topology data determined for the random position of the user being somewhere between 0 to 5 km. The x-axis of the chart shows the different geocoding query types (i.e., full address, street and house

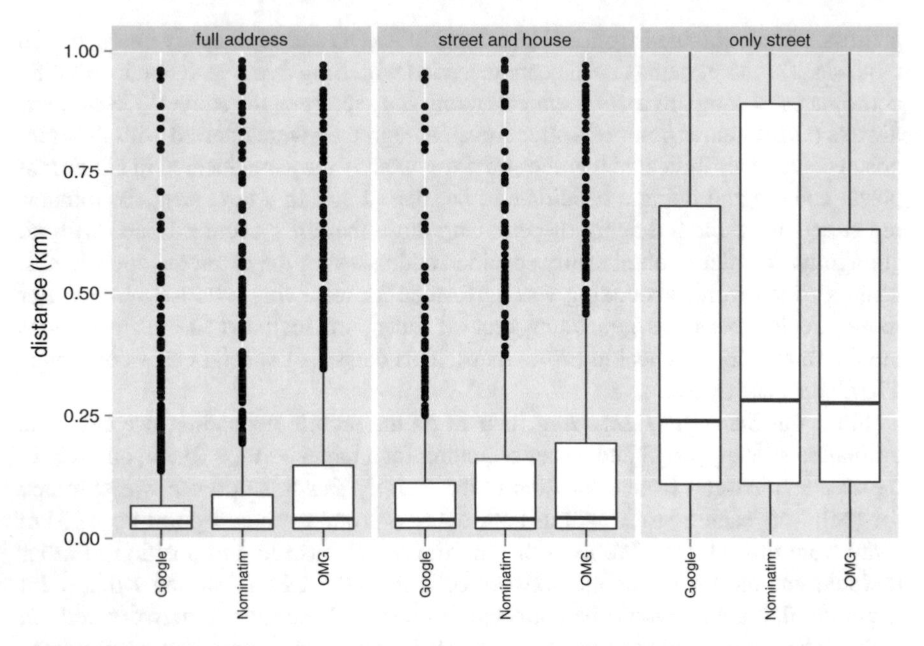

Fig. 10.6 Distance of geocoding result to target address for the random dataset with uniquely available addresses

number, and only street name) and geocoding services, whereas the y-axis highlights the distribution of the distances calculated in [km] in the form of box plots.

Table 10.1 lists all relevant median values for the different geocoding query types and geocoding services. Here, we can see that for the normal dataset of randomly generated addresses, the results are more or less even with slight differences. We cannot outperform *Nominatim* or *Google* when assembling geocoding queries with full or partial address information indicating that network topology data has no (positive) effect on the results when the address information is more or less completely available. Even though our OMNG dataset has been initially extracted from OSM and therefore the same geocoding performance as *Nominatim* could be expected, our results are slightly worse than *Nominatim*. This is due to the fact that in contrast to geocoding services running in production, we did not focus on implementing

Table 10.1 Median values for distances of geocoding results to target addresses using the random dataset

	Google	Nominatim	OMNG
Full address	0.033530	0.034984	0.040206
Partial address	0.03865	0.03996	0.04245
Street-level address	0.240	0.28233	0.27695

additional and sophisticated algorithms to match addresses that contain typos, etc. We rather solely concentrated on enriching geocoding requests with network data in order to observe its effects on the results.

Having a look at the results of queries for only street names, however, we can see that we have performed slightly better than *Nominatim* showing that network topology data can improve geocoding results. The slight improvement is mainly achieved for the few ambiguous addresses (existing more than once in Berlin) that are available within the randomly created dataset.

In Table 10.2, we listed the percentage of "correct" geocoding results defined as distances being below 500 m to the target addresses. Here, we can observe that for full address information, *Google* determines almost 96% of the addresses correctly, whereas *Nominatim* achieves 92% correct results both being better than OMNG. However, for partial addresses and only street names, we achieve better results than our competitors with 84% and 68% correctness respectively.

Table 10.2 Percentage of correct geocoding results (defined as being below 500 m) using the random dataset

	Google	Nominatim	OMNG
Full address	0.9578415	0.9198988	0.886172
Partial address	0.8381113	0.811973	0.841484
Street-level address	0.6610455	0.6172007	0.6795953

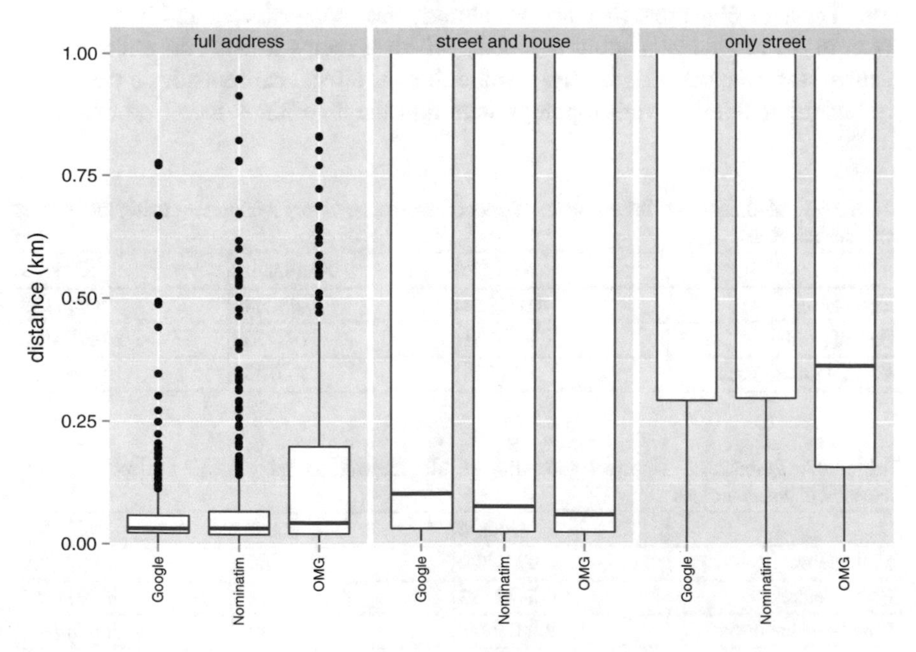

Fig. 10.7 Distance of geocoding result to target address for the dataset with special cases

Figure 10.7 shows the results achieved for the dataset of special address cases, i.e., ambiguous addresses existing several times in the city. This chart highlights that our approach serves the desired use case well. The *OMN Geocoder* clearly outperforms *Nominatim* and *Google* except when supplied with a full address indicating that mobile network topology data can improve geocoding services to a great extent. Having a random (near) distance between 0 to 5 km to the desired target addresses, the chance is greater that the target addresses are covered by the serving cell or the corresponding LA.

In Table 10.3, the median values for the different geocoding request types and services are listed based on the dataset of special address cases. There is a slight difference in the median values when supplying full addresses to the geocoding request confirming again that mobile network data has no effect when address information is given completely. However, the values for partial addresses (e.g., `0.10281` vs. `0.060112`) and street-level addresses (e.g., `3.27026` vs. `0.364523`) clearly display that cell data can improve the geocoding results to a great extent.

Table 10.4 gives an overview about the percentage of "correct" geocoding results defined as distances being below 500 m to the target addresses using the dataset for special address cases. Here, we can also see that for partial and street-level addresses, we perform best with 69% and 55% correctness respectively. In contrast to Table 10.2, we can further observe that there is a huge gap between the results showing that cell data plays a major role for improving geocoding.

Summarizing the evaluation results, we can state that *Google* and *Nominatim* deliver very good geocoding results for complete and unique address information. The algorithms that they use are already very sophisticated and hard to beat. Even in the mobile geocoding use case, cell data seems to have no effect on the results. However, we need to keep in mind that the effect also depends on the "quality" of the mobile network topology data and the algorithms based on this data.

Table 10.3 Median values for distances of geocoding results to target addresses using the dataset with special cases

	Google	Nominatim	OMNG
Full address	0.031824	0.032385	0.044135
Partial address	0.10281	0.07702	0.060112
Street-level address	3.27026	2.07033	0.364523

Table 10.4 Percentage of correct geocoding results (defined as being below 500 m) using the dataset with special cases

	Google	Nominatim	OMNG
Full address	0.9747899	0.9243697	0.8571429
Partial address	0.5483193	0.5777311	0.6911765
Street-level address	0.3319328	0.3487395	0.5588235

The *OpenMobileNetwork* is based on crowdsourcing, which implies that the data could be error-prone to some degree and is not as accurate as real network topology data of an operator. In addition, we have rather used a simple algorithm in our first implementation comparing the target address with the cell and LA the user is connected to. We assume that, for example, integrating information about neighboring cells into the geocoding algorithm and having more accurate information about the coverage areas of mobile network cells would improve the geocoding results even when supplying full addresses.

The evaluation of the geocoding results for ambiguous address information, on the other hand, clearly showed the positive effect of cell data in comparison to the results of our competitors. We distinctly outperformed *Google* and *Nominatim* for geocoding queries using parts of an address or only the street name indicating that mobile network data could improve the preciseness of geocoding results for addresses that exist several times in a city.

Part IV
Conclusion

Chapter 11
Conclusion

This thesis outlined the evolution of mobile network operators from bit pipes to service providers by exploiting their asset - the mobile network data. It highlighted the restrictions and drawbacks of other solutions providing services on top of the network information and added a new dimension to the discussion by proposing a semantic modeling and enrichment of mobile network and WiFi access point topology data according to the principles of *Linked Data* in combination with interlinks to diverse context sources within the *LOD Cloud*.

In general, this work tried to find an answer to the research questions formulated in Sect. 1.1:

1. How to model an ontology that incorporates a geographic and topological view on mobile as well as WiFi networks and further takes dynamic network context information into consideration?
2. How to collect and accurately estimate worldwide mobile and WiFi network data that suffices the requirements for context-aware services in the In-house, B2C as well as B2B application areas?
3. How to interlink the semantically enriched network topology data to diverse context sources?
4. How to create a dataset and its corresponding vocabulary that extends static location data with dynamic context information?
5. How to enable network operators to leverage semantically enriched mobile and WiFi network data for providing innovative context-aware services?

A. Uzun, *Semantic Modeling and Enrichment of Mobile and WiFi Network Data*, T-Labs Series in Telecommunication Services,
https://doi.org/10.1007/978-3-319-90769-7_11

11.1 Summary of the Contribution

For this purpose, we first set the stage for further work in Chap. 3 by bringing together the requirements for a semantically enriched mobile and WiFi network data platform. Here, we gave an overview about the In-house, B2C, and B2B application areas in which telecommunication providers are potentially able to offer semantically enriched services and consolidated a list of context data requirements with a special focus on the power management scenario and the network optimization use cases described in Sect. 7.1. Furthermore, we made a deep dive into the *LOD Cloud* and analyzed its applicability as a third-party context provider in terms of discoverability, availability, and contents of the datasets. We identified context data to be of particular interest for our three application areas that are already available in a structured form and came to the conclusion that the interlinking nature of *Linked Data* provides great potential for improving the QoE of context-aware services in the telecommunications domain. Moreover, we listed the functional as well as non-functional requirements for a platform offering semantically enriched network topology data with interlinks to other datasets of the *LOD Cloud*.

In the next step, we shed light on the concept and design process of semantically modeling mobile as well as WiFi network data in Chap. 4 and presented the *OpenMobileNetwork* as the core contribution of the doctoral thesis, which is a platform for providing this data based on the principles of *Linked Data*. Here, we gave an overview of several data sources already offering mobile network as well as WiFi AP information (see Sect. 4.1) and decided to build our own dataset due to access restrictions, unavailable network context parameters, or missing quality evaluations of the available data sources. We further designed a functional architecture for the *OpenMobileNetwork* by having a detailed discussion on the advantages and drawbacks of several architectural alternatives such as a central crowdsourcing platform, an architecture for interlinking operator-specific data on a meta level, or a data federation approach. We identified the crowdsourcing platform as an applicable form for our functional architecture and utilized various facets of crowdsourcing methods, such as systematic warwalking and wardriving, a gamification approach, or a background crowdsourcing service, for maximizing the potential of collecting network context data in the form of network measurements. Later, we illustrated how network topologies are calculated out of a set of crowdsourced network measurements based on a discussion of various position and coverage area estimation algorithms.

After finishing the preliminary steps, we described the procedure of semantifying network context data by modeling the *OpenMobileNetwork Ontology* that consists of a set of static and dynamic network context ontology facets. This ontology represents mobile networks and WiFi APs from a topological perspective and geographically relates the coverage areas of these network components to each other. The chapter is concluded with a coarse explanation of the instance data triplification process and an illustration of the *OMN VoID Description* publishing meta data about the *OpenMobileNetwork* dataset.

Having the *OpenMobileNetwork Ontology* and the corresponding dataset ready, we focused on the approach of interlinking diverse context sources to the network topology data in Chap. 5. For this purpose, we first interlinked the OMN to available datasets in the *LOD Cloud* based on our geographic mapping model and utilized *LinkedGeoData* as a context source for interconnecting geo-related data, whereas descriptions in *DBpedia* were used to relate various (textual) representations for mobile network information such as operator names.

Taking the limitations of the available geo-related datasets as a reference, we further presented *Linked Crowdsourced Data* as another contribution of this thesis, which is a crowdsourced dataset linking dynamic parameters, specific context situations as well as additional domain-specific information to static location data allowing the development of semantic location analytics services. Here, we also discussed the crowdsourcing approach of collecting and processing third-party context information and made a deep dive into the modeling of the corresponding *Context Data Cloud Ontology*. In addition, we also introduced the *OpenMobileNetwork Geocoding Dataset* as another context source for interlinking address data to the *OpenMobileNetwork* that enables address-related services such as geocoding.

After finishing the conceptual work, we gave an insight into the implementation of the *OpenMobileNetwork* in Chap. 6 by providing an end-to-end view on all platform components including the system architecture, the smartphone clients for network context data collection, the backend server as well as the *OpenMobileNetwork* website with its various coverage visualization maps.

In order to ultimately find an answer to the fifth research question, Chap. 7 introduced In-house, B2C, and B2B services that were designed as proofs of concepts highlighting the added value of the *OpenMobileNetwork*. The In-house service, for example, consisted of a power management scenario that enables operators to optimize their networks by de- and reactivating mobile network cells based on the capacity demand. Several *Semantic Positioning* solutions, on the other hand, were presented as B2C services that improve classic geocoding as well as geofencing methods and add semantic features to proactive LBSs, while the usage of semantically enriched location analytics was illustrated as an example for a B2B service.

The corresponding proof of concept service implementations, such as the *OpenMobileNetwork for ComGreen Demo*, the *Friend Tracker* and *Popular Places Finder* services, the *OpenMobileNetwork Geocoder* as well as the *Location Analytics Map*, were demonstrated in Chap. 8.

Due to the fact that the performance of the provided services (in terms of accuracy and user experience) closely depends on the preciseness of the approximated network topologies and hence an accurate geographic mapping of them, a quality analysis of the *OpenMobileNetwork* dataset was conducted in Chap. 9 based on a distance comparison of approximated mobile network cells within the OMN to the real positions of these cells given by an operator. We applied this distance comparison twice within a time frame of approximately 1.5 years in order to observe if the estimations led to better results with more collected network measurements. The results indicated that the increasing number of network measurements over time had a positive impact on the network topology estimations and thus proved that the semantically enriched services based on the geographic mapping model of the *OpenMobileNetwork* rely on a solid and accurate network topology dataset.

Based on the quality analysis of our data, we analyzed the applicability of our proposed services in a second step by performing in-depth accuracy evaluations for the *Semantic Tracking* as well as the *Semantic Geocoding* respectively.

The accuracy of the *Semantic Tracking* function was evaluated in terms of a distance calculation between the tracked user's current position at the moment the tracking party gets notified by an *entering* or *leaving* event and the position of the location entity itself within LCD. After consolidating the results of several evaluation tours, we showed that our *Semantic Tracking* approach delivered competitive results and that semantically enriched mobile and WiFi network topology data in combination with links to geo-related datasets can be a good alternative to classic geofencing methods in areas with dense network coverage (e.g., cities). We further proposed *semantic geofences* in combination with network topologies as a hybrid solution for rural areas.

The performance of our *Semantic Geocoding* function, on the other hand, was compared against its competitors in terms of ambiguous or incomplete address input data. In contrast to the achieved geocoding results for complete and unique address information where mobile network cell data seemed to have no effect on the output, the results for ambiguous address data clearly showed the positive effect of the network topology data by outperforming the competitors for geocoding queries using parts of an address or only the street name.

11.2 Discussion of the Research Results

Turning back to the initial research questions, we strongly believe that the fusion of mobile and WiFi network data with semantic technologies as well as diverse context sources is a very unique approach that is not considered yet (in its full dimension) and that provides great potential for establishing telecommunication providers as service enablers.

The *OpenMobileNetwork* as a proof of concept platform implementation highlights the flexibility of applying semantically enriched network topology data in various application areas and showcases the added value of semantic data representation. It enables sophisticated semantic queries that do not (only) rely on network parameters, geofences, or geo coordinates, but rather on semantic relations between mobile network cells, WiFi APs, and external data sources (e.g., POIs) as illustrated in the service examples in Chaps. 7 and 8. Moreover, the uniform RDF data format eases the interlinking of new context data sources and the retrieval of data for potential context-aware service developers.

Linked Crowdsourced Data supports the dataset of the *OpenMobileNetwork* by providing the necessary context data richness in order to realize fine-grained and semantically enriched location analytics services. It elevates geo-related datasets from being rather of static nature to also publish dynamic context information related to locations and represents a blueprint (for telecommunication providers) of how third-party context information could be modeled in *Linked Data* format.

Taking the evaluation results as a reference, we can see our initial assumption being verified that mobile network operators are able to exploit semantically enriched network topology data for providing innovative services that might distinguish them from other service providers or that might perform better (in some aspects) as the services of the competitors. The verification will be even stronger if we think of utilizing real and accurate network topology data as well as applying professional and scalable app development.

Also from the perspective of the *Web of Data*, we believe that the datasets of the *OpenMobileNetwork* as well as *Linked Crowdsourced Data* are valuable contributions to the LOD community with content that was not available before within the *LOD Cloud*. These datasets can also be used in a standalone fashion independent from the perspective of a mobile network operator. The advantage of these datasets is that they are not dependent on any research project and that they are constantly extendable via crowdsourcing (since the necessary apps are available at *Google Play*) by anyone who is interested in exploiting them for his own purposes.

However, the expressiveness of the *OpenMobileNetwork Ontology* as well as the *Context Data Cloud Ontology* is limited at the moment since we rather focused on providing semantically enriched network topology data and interlinking this information with heterogeneous data sources within the *LOD Cloud*. In order to further extend the added value of applying semantic technologies to mobile network data, more expressiveness and semantics should be integrated into the ontologies out of which implicit knowledge is generated using powerful reasoning tools (see footnote 5 in Chap. 1).

Furthermore, using *Linked Open Data* for commercial purposes is still an open problem due to licensing issues [74] as well as missing data quality control mechanisms and service level agreements. These aspects need to be tackled also if a production release of such a platform is planned by an operator.

All in all, the foundation for the usage of semantically enriched network topology data is set. There is a variety of potential research work that can be applied on top of this contribution. Details are discussed in Chap. 12.

Chapter 12
Future Outlook

In the course of time, we identified several areas, in which we could potentially extend our research. Section 12.1 discusses approaches of how to add more semantic information to the *OpenMobileNetwork* and extend the expressiveness of the ontology, while Sect. 12.2 proposes a *Location Analytics Framework* for identifying trends and providing valuable analytics information to third-party developers and business owners. In Sect. 12.3, on the other hand, we introduce the first steps of the conceptual work towards a context data discovery within the *LOD Cloud*.

12.1 Adding More Semantic Information to the OpenMobileNetwork

In our current work, we focused on semantically modeling mobile and WiFi network data from a topological perspective as a foundation for enabling semantically enriched services on top of the data. However, there is potential in integrating much more semantic information to the *OpenMobileNetwork* and hence extending the expressiveness of the corresponding ontology by abstracting the network topology data from a geospatial level (with geo coordinates for positions and polygons) to a fully-fledged semantic level, where only interdependencies between cells, third-party context data (e.g., POIs), and region-specific information are described via RDF statements (as partially also proposed in Sect. 4.2.1).

One straightforward example could be to interlink district, city, and country names (coming from *DBpedia* or *GeoNames*, for instance) with the network cells in order to optimize SPARQL requests by avoiding the search for locations with the `bif:st_distance` function. Such a relation would improve the performance of

A. Uzun, *Semantic Modeling and Enrichment of Mobile and WiFi Network Data*, T-Labs Series in Telecommunication Services,
https://doi.org/10.1007/978-3-319-90769-7_12

the retrieval process and would facilitate the formulation of queries by just asking for "all cells that cover the Schöneberg district in Berlin, Germany", for example.

Another interesting approach could be to define an additional *OpenMobileNetwork Region Ontology* facet that describes region-specific information. This ontology could incorporate static societal information such as a certain district in Berlin being well-known for its Turkish and Arab community. Based on powerful reasoners, it could generate implicit knowledge by inferring that *Turkish and Arab people are mostly Muslim who mainly eat Halal food* meaning that *pork is not allowed*. In addition, dynamic changes in the district could be (automatically) integrated into the ontology (and its corresponding dataset) by performing long-term location analytics (based on *Linked Crowdsourced Data*) such as information that more and more people tend to *listen to rock music*. If this information was attached to the mobile network cells and WiFi APs covering this district, sophisticated semantic location analytics services could be provided that take this kind of information into consideration.

The *OpenMobileNetwork Region Ontology* could also be utilized for geofences. As mentioned in Sect. 10.1, our *Semantic Tracking* service works well in areas with dense network coverage such as in cities. In rural areas, however, the coverage areas of cells are much bigger and WiFi APs are not distributed as much as in cities meaning that the network topology will not be sufficient to replace fine-grained geofences. For rural areas, we therefore propose a hybrid solution combining semantically enriched geofences and network topologies.

An automatically created circular geofence for a location entity is already represented by the *Location* facet of the *Context Data Cloud Ontology* that defines the `cdc-owl:radius` predicate for a location entity as a description of its ground area (in meters). A more sophisticated model of a polygonal geofence could be created using the `ogc:asWKT` and `sf:LineString` properties. Both types of a semantic geofence could be linked to the information described by the *OpenMobileNetwork Region Ontology*.

Adding more semantic information and extending the expressiveness of the *OpenMobileNetwork Ontology* as well as the *Context Data Cloud Ontology* will highlight the added value of semantic technologies to a greater extent.

12.2 Location Analytics Framework based on the OpenMobileNetwork

Section 8.4 demonstrated the *Location Analytics Map* that visualizes aggregated and rudimentally analyzed LCD data on a map. We propose to extend this work by a *Location Analytics Framework* as part of the *OpenMobileNetwork* that provides well-defined interfaces and visualization options with which a mobile network operator becomes capable of providing various location analytics information to third-party developers and business owners.

This framework needs to provide insight about the activities within different mobile network regions, which could be achieved by a correlation of several data sources. Assuming that privacy issues are handled, the raw network measurement data of the *OpenMobileNetwork* makes it already possible to infer movements and stationary phases of users as exemplary shown in [C4]. For this purpose, sophisticated data mining tools, such as *Weka*,[1] in combination with semantic trajectory modeling concepts [111] could be utilized in order to derive user movements and correlate this data with the network topology information of the *OpenMobileNetwork* as well as context information available within *Linked Crowdsourced Data* and other external data sources. In addition, semantic user profiles (e.g., preferences, demographic information, or favorite locations) as defined by the *User Profile* facet of the *Context Data Cloud Ontology* could be integrated into the correlation process.

The fusion of these data sources will make the *Location Analytics Framework* capable of identifying trends as they are coming up and will provide valuable information to business owners such as a statistical count of vegetarians passing by their restaurant.

12.3 Context Data Discovery in the LOD Cloud

In Sect. 3.2.1, we have shown the potential of the *LOD Cloud* to be exploited as a valuable third-party context source and interlinked several datasets to the *OpenMobileNetwork* (see Chap. 5) as a foundation for a variety of context-aware services (see Chap. 7). However, enriching services with more context data from the *LOD Cloud* requires a lot of manual effort in finding appropriate datasets as well as the corresponding SPARQL endpoints and in understanding the schema behind them. This is very time-consuming and often not applicable when developing new services for the network operator, which is why dataset and context information discovery is a crucial factor when dealing with the *LOD Cloud*.

For this purpose, we propose a *Context Data Lookup Service* [J2] as a main future outlook to the contribution of this thesis that enables context data discovery within the *LOD Cloud*. It consists of a new *Context Meta Ontology (CMO)*[2] (see Sect. 12.3.1) that provides a meta description for available context data within a dataset and a *Context Meta Ontology Directory (CMOD)* (see Sect. 12.3.2) for registering the CMOs related to the datasets and making them available for context data discovery. By classifying major vocabulary concepts to context categories using the CMO, the *Context Data Lookup Service* facilitates search and discovery of appropriate concepts and underlying instance data for a given context situation.

---

[1] http://www.cs.waikato.ac.nz/ml/weka/.

[2] *cmo*, http://www.contextdatacloud.org/cmo/ontology/.

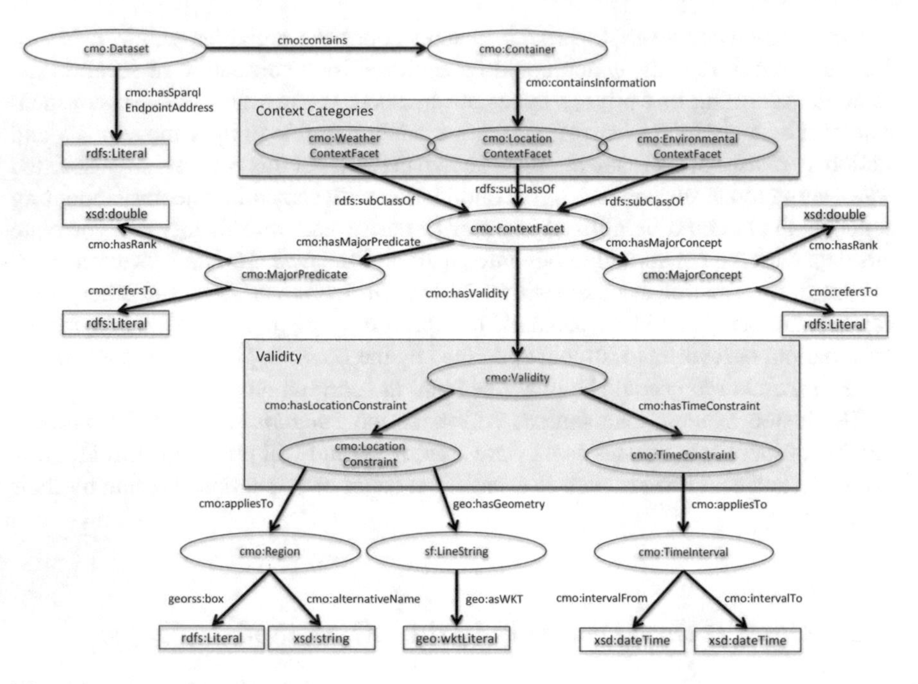

Fig. 12.1 Context Meta Ontology

12.3.1 Context Meta Ontology

Inspired by VoID [162], the *Context Meta Ontology* provides a meta description for relevant context information contained within a dataset. It needs to be populated with information by each dataset owner in order to make their dataset available within the *Context Data Lookup Service*. The CMO is based on RDFS [58] as well as OWL [137] and is identified by the namespace http://www.contextdatacloud.org/cmo/ontology/ with the prefix `cmo`. Resources, on the other hand, are represented by http://www.contextdatacloud.org/cmo/resource/ using the prefix `cmor`. When applying the ontology to a certain dataset, we recommend to use a standardized namespace in the form of http://URI/cmo/. Figure 12.1 shows parts of the CMO.

In order to enable search and discoverability of the context information within a dataset, the predicate `cmo:hasSparqlEndpointAddress` relates the `cmo:Dataset` to the URI of the corresponding SPARQL endpoint. The ontology further classifies context according to high-level categories based on Sect. 3.2.1.2. These categories are represented as predefined `cmo:ContextFacets` (e.g., `cmo:LocationContextFacet`, `cmo:WeatherContextFacet`, or `cmo:EnvironmentalContextFacet`), which can easily be extended as required. Since datasets may contain information relating to more than a single context facet (e.g., location and weather), datasets can be partitioned into several `cmo:Containers` holding different contextual information (`cmo:contains`

Information). For each cmo:Container, the individual type of contextual information is specified using the above mentioned context categories.

As further described in Sect. 3.2.1.2, our content approximation method identifies the most often used properties and classes within a dataset. This information is of high relevance for context-aware service developers, so that they know what is inside a dataset and which concepts mainly to utilize in order to enrich their services with relevant data. Our CMO incorporates this information by using the cmo:MajorConcept and cmo:MajorPredicate concepts. A cmo:MajorPredicate cmo:refersTo the URI of a popular predicate within a dataset and ranks it (cmo:hasRank) according to its weight (calculated by its count). cmo:MajorConcept, on the other hand, represents and ranks popular classes. However, due to the fact that ontology concepts within the *LOD Cloud* are not always self-explanatory as it is with most of the concepts used in *LinkedGeoData* (e.g., lgdo:Amenity or geom:Geometry), a mapping of the popular concepts to the provided cmo:ContextFacets is required in order to enable a discovery of ontology concepts for a given context. In the first version of the CMO, we assume that dataset owners manually categorize their concepts. Even though this approach is time-consuming, it provides the highest accuracy and also takes the meaning of concepts into consideration. Furthermore, the CMO only supports the mapping of major concepts to context facets without taking the schematic relations into consideration. After the automatic discovery of location context data, for example, a service developer still needs to have a final look at the schema in order to know how to exploit the instance data. In future, the relevant schema parts should also be classified to context facets based on the schema approximation techniques introduced in [C12, J2] in order to ease the usage process for developers.

To specify the validity of the respective contextual information contained within a dataset, the concept of cmo:Validity is used, which augments a specific cmo:ContextFacet with validity constraints. cmo:Validity indicates whether a certain container of contextual information can be used as a context provider in a certain situation. In this regard, it is differentiated between *local* (cmo:LocationConstraint) and *temporal* (cmo:TimeConstraint) validity.

cmo:LocationConstraint allows the filtering of data according to specific locations. For instance, the dataset of the *OpenMobileNetwork* serves mobile and WiFi network topology information that is bound to certain locations (given as WGS84 coordinates). As this dataset does not contain information for all regions of the world, the cmo:LocationConstraint is set to specify the limited applicability with respect to location. Here, we can set two types of cmo:Location Constraints: A cmo:Region is a textual representation of a region (e.g., a city or a district of a city) whose area is described with geospatial data (e.g., georss:box). One example could be a boundary box describing the region of Berlin that enables search requests for location context data (e.g., POIs) only available in that area. Interlinks to other datasets providing a representation of a region with geospatial data is also possible. The other type of cmo:LocationConstraint describes a geospatial area (e.g., a sf:LineString) independent from its

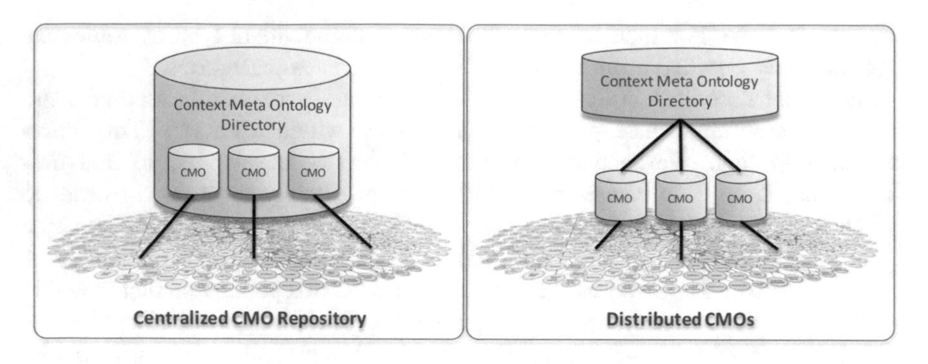

Fig. 12.2 Context Meta Ontology Directory – alternative architectures

textual representation using the *OGC GeoSPARQL Vocabulary*. The `cmo:TimeConstraint` concept, on the other hand, filters data according to time constraints, e.g., only up-to-date sensor data. This constraint is based on a temporal description within the respective dataset using `xsd:dateTime`. As time is only a single dimension, the applicability can be defined as an interval (`cmo:TimeInterval`) and a certain point in time needs to be included in the interval in order to be valid or applicable.

Validity constraints can also be combined defining only data for a specific region and available within a certain time interval, for example. The list of validity constraints is not considered as complete and can be extended as required.

12.3.2 Context Meta Ontology Directory

The *Context Meta Ontology Directory*[3] is a central repository that keeps track of all CMOs and that is the first entry point for context-aware service developers when searching for context data. We propose two architectural alternatives for the interworking between the CMOD and the CMOs, which can be seen in Fig. 12.2.

The *Centralized CMO Repository* is basically one triplestore based on the CMO schema that maintains all meta descriptions of datasets. This architecture is easier to maintain for the CMOD provider, but has the drawback that dataset owners need to upload their meta description to this central repository, which makes later updates and changes to the CMO more difficult. Furthermore, dataset owners "lose control" over their meta description that is also not desirable.

The *Distributed CMO* architecture, on the other hand, enables dataset owners to keep their CMO on their own server. The CMOD only maintains a link to the CMO and controls its availability. This is depicted in Fig. 12.3.

[3]http://cmod.contextdatacloud.org/.

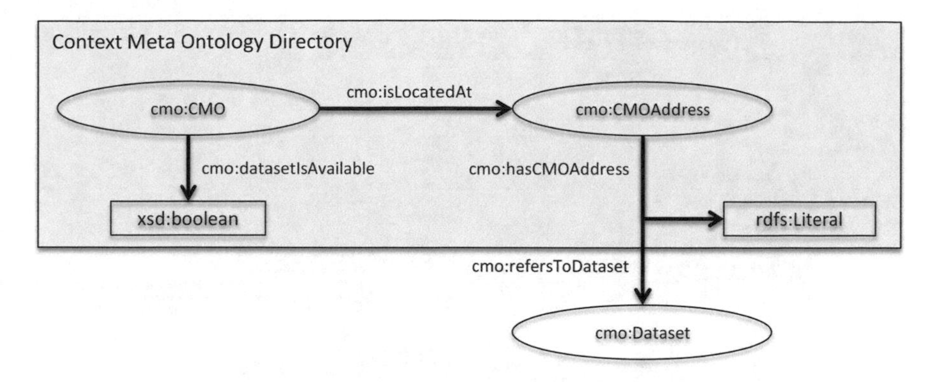

Fig. 12.3 Context Meta Ontology Directory – interlink to distributed dataset CMO

`cmo:CMO` is a concept representing a CMO for a certain dataset, which `cmo:isLocatedAt` a certain `cmo:CMOAddress`. This `cmo:CMOAddress` refers to (`cmo:refersToDataset`) the meta description of the `cmo:Dataset`. The availability of the dataset endpoint is checked frequently with a cronjob and set with a boolean value (`cmo:datasetIsAvailable`). This predicate is linked on purpose to `cmo:CMO` rather than to `cmo:Dataset` in order to make sure that the CMO of the dataset is not considered in the dataset discovery process if the availability is `0`.

The first version of our *Context Data Lookup Service* and the example discussed in Sect. 12.3.3 is based on the centralized architecture. However, our vision is to provide a distributed solution whenever the CMO reaches a certain popularity within the LOD community.

12.3.3 Querying the Context Meta Ontology Directory

In order to showcase the applicability of the *Context Data Lookup Service*, an exemplary CMO is created for *LinkedGeoData* (see Fig. 12.4). The `cmor:LocationContextFacet_1` highlights two major concepts (`lgdo:Amenity` and `lgdm:Node`) and one predicate (`geom:geometry`) with their respective weights. Furthermore, a `cmor:LocationConstraint_1` is set with Berlin as a region (`cmor:Berlin`) and its boundary box.

We assume that a context-aware service developer intends to build a service. He does not know what kind of data is available within the *LOD Cloud* that he could use for his application. To get an overview about datasets that are of interest with respect to his planned application, he queries the *CMO Directory* for all datasets and covered context facets. The corresponding SPARQL query in Listing 12.1 returns all specific datasets including their endpoint addresses as well as the contextual information facets covered by them. Applied to the *LinkedGeoData*

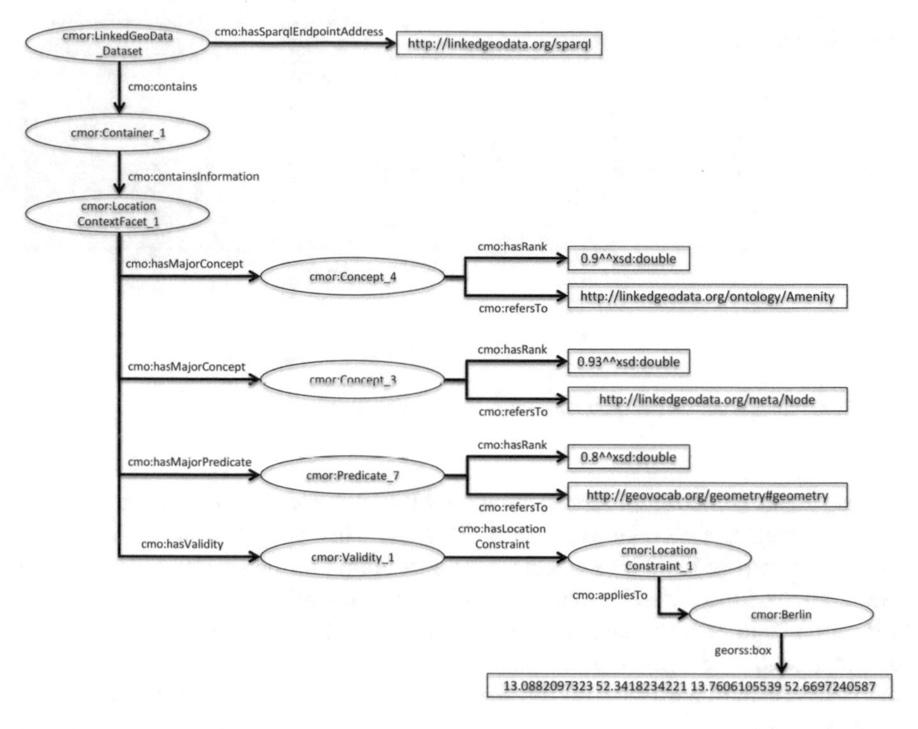

Fig. 12.4 Exemplary Context Meta Ontology for LinkedGeoData

CMO example, the system returns `cmor:LinkedGeoData_Dataset` including its SPARQL endpoint (http://linkedgeodata.org/sparql) as a dataset and the `cmor:LocationContextFacet_1` as a context facet.

```
SELECT DISTINCT ?dataset ?address ?facet
WHERE {
  ?dataset a cmo:Dataset .
  ?dataset cmo:hasSparqlEndpointAddress ?address .
  ?dataset cmo:contains ?container .
  ?container a cmo:Container .
  ?container cmo:containsInformation ?facet .
  ?facet a ?contextFacet .
  ?contextFacet rdfs:subClassOf cmo:ContextFacet .
}
```

Listing 12.1 SPARQL query to retrieve available datasets and covered context facets

After having received an overview of the data, he needs to identify the context facets for which data can be found that is applicable with respect to the location and time the developer is interested in. Here, we assume that the location of a specific user (group) is given. In case a service is required to function worldwide, *rules* need to be specified that forward a location-dependent request to the appropriate dataset for enrichment or aggregation of relevant background information. Listing 12.2 shows a SPARQL query that retrieves location constraints for a given dataset's context facet that are specified for regions. For this purpose, the caller passes two arguments, namely the dataset (`$DATASET$`) and the context facet (`$FACET$`) he is interested in, and gets the location constraints as a set of regions and bounding boxes. In our CMO example, the system returns `cmor:Berlin` and its boundary box "`13.0882097323 52.3418234221 13.7606105539 52.6697240587`".

```
SELECT DISTINCT ?region ?location
WHERE {
  $DATASET$ a cmo:Dataset .
  $DATASET$ cmo:contains ?container .
  ?container a cmo:Container .
  ?container cmo:containsInformation $FACET$ .
  $FACET$ cmo:hasValidity ?validity .
  ?validity cmo:hasLocationConstraint ?constraint .
  ?constraint cmo:appliesTo ?region .
  ?region georss:box ?location .
}
```

Listing 12.2 SPARQL query to retrieve location constraints for a given dataset's context facet

In a third step, the developer can query the CMO for *major predicates* and *classes*, which ease the process of constructing SPARQL queries according to the dataset's specific vocabulary. Listing 12.3 lists all major concepts including their rank for a given dataset's context facet. For this purpose, the caller again needs to pass the dataset (`$DATASET$`) and the specific context facet (`$FACET$`) he is interested in. The result provides insight to the developer what kind of concepts are used to describe the contextual information of the given facet in the respective dataset. For our *LinkedGeoData CMO* example, the system returns http://linkedgeodata.org/ontology/Amenity as well as http://linkedgeodata.org/meta/Node for the `cmor:LocationContextFacet_1` with `0.9` and `0.93` as their respective weights. Looking at this result set and checking their schematic relations within *LinkedGeoData*, the developer realizes that this dataset provides *amenities* (i.e., points of interest) in Berlin, which he can integrate into his location-based application.

```
SELECT DISTINCT ?concept ?rank
WHERE {
  $DATASET$ a cmo:Dataset .
  $DATASET$ cmo:contains ?container .
  ?container a cmo:Container .
  ?container cmo:containsInformation $FACET$ .
  $FACET$ cmo:hasMajorConcept ?majorconcept .
  ?majorconcept cmo:refersTo ?concept .
  ?majorconcept cmo:hasRank ?rank .
}
ORDER BY DESC(?rank)
```

Listing 12.3 SPARQL query to retrieve all major concepts for a given dataset's context facet

In case we have a knowledge base containing the descriptions of several datasets according to the *Context Meta Ontology*, developers of semantically enriched context-aware services will be able to search for datasets providing diverse contextual information that can be integrated into their service. This is done by classifying the major classes and predicates of a dataset into context facets in combination with validity constraints. Knowing the context facets to be used within a service and the dataset concepts representing these context categories, the process of constructing SPARQL queries according to the dataset's specific vocabulary for retrieving the instance data will be facilitated.

In future, the CMO needs to be extended by also taking relevant parts of the schema into consideration when classifying concepts to context categories, so that the developer knows directly how to use the instance data of a dataset without the need of checking the dataset's schema manually. Furthermore, a direct connection between the CMO and dataset resources would be desirable enabling the retrieval of instance data from the *LOD Cloud* through the CMO by totally avoiding a priori knowledge about SPARQL endpoints and vocabulary concepts.

References

1. Ericsson mobility report - on the pulse of the networked society (2012). White Paper. http://hugin.info/1061/R/1659597/537300.pdf
2. Abele A, McCrae JP, Buitelaar P, Jentzsch A, Cyganiak R (2017) The Linking Open Data cloud diagram. http://lod-cloud.net/
3. Adomavicius G, Tuzhilin A (2011) Context-aware recommender systems. Recommender systems handbook. Springer, US, pp 217–253
4. Alexander K, Cyganiak R, Hausenblas M, Zhao J (2009) Describing linked datasets - on the design and usage of voiD, the "Vocabulary of Interlinked Datasets". In: WWW 2009 workshop: linked data on the web, LDOW '09, Madrid, Spain
5. Alt F, Shirazi AS, Schmidt A, Kramer U, Nawaz Z (2010) Location-based Crowdsourcing: extending crowdsourcing to the real world. In: Proceedings of the 6th Nordic conference on human-computer interaction: extending boundaries, NordiCHI '10. ACM, New York, pp 13–22
6. Atemezing GA, Hyland B, Villazón-Terrazas B (2014) Best practices for publishing linked data. W3C note, W3C. https://www.w3.org/TR/ld-bp/
7. Attard J, Scerri S, Rivera I, Handschuh S (2013) Ontology-based situation recognition for context-aware systems. In: Proceedings of the 9th international conference on semantic systems, I-SEMANTICS '13. ACM, New York, pp 113–120
8. Bandara A, Payne T, Roure DD, Clemo G (2004) An ontological framework for semantic description of devices. In: Proceedings of the 3rd international semantic web conference, Hiroshima, Japan
9. Bareth U, Küpper A (2011) Energy-efficient position tracking in proactive location-based services for smartphone environments. In: Proceedings of the IEEE 35th annual computer software and applications conference, COMPSAC '11, Los Alamitos, CA, USA. IEEE Computer Society, pp 516–521
10. Bareth U, Küpper A, Ruppel P (2010) geoXmart - a marketplace for geofence-based mobile services. In: Proceedings of the IEEE 34th annual computer software and applications conference, COMPSAC '10. IEEE, pp 101–106
11. Barnaghi P, Presser M (2010) Publishing linked sensor data. In: Proceedings of the 3rd international workshop on semantic sensor networks, SSN '10, Shanghai, China
12. Barraclough C (2011) The value of 'smart' pipes to mobile network operators. Strategy Research, STL Partners/Telco 2.0

A. Uzun, *Semantic Modeling and Enrichment of Mobile and WiFi Network Data*, T-Labs Series in Telecommunication Services,
https://doi.org/10.1007/978-3-319-90769-7

13. Becker C, Bizer C (2009) Exploring the Geospatial Semantic Web with DBpedia Mobile. Web Semant Sci Serv Agents World Wide Web 7(4):278–286
14. Bellavista P, Corradi A, Fanelli M, Foschini L (2012) A survey of context data distribution for mobile ubiquitous systems. ACM Comput Surv 44(4):24:1–24:45
15. Benjamins VR (2014) Big data: from hype to reality? In: Proceedings of the 4th international conference on web intelligence, mining and semantics, WIMS '14. ACM, New York, pp 2:1–2:2
16. Bergmann T, Bunk S, Eschrig J, Hentschel C, Knuth M, Sack H, Schüler R (2013) Generating a linked soccer dataset. In: Proceedings of the 9th international conference on semantic systems, I-SEMANTICS '13. ACM, New York, pp 146–149
17. Bernardin P, Yee M, Ellis T (1997) Estimating the range to the cell edge from signal strength measurements. In: IEEE 47th vehicular technology conference, vol 1, pp 266–270
18. Berners-Lee T (2000) Weaving the web: the past, present and future of the World Wide Web by its inventor. Texere Publishing, London
19. Berners-Lee T (2008) Linked open data. In: Linked data planet. https://www.w3.org/2008/Talks/0617-lod-tbl/
20. Berners-Lee T (2009) Linked data. Personal note. https://www.w3.org/DesignIssues/LinkedData.html
21. Berners-Lee T, Hendler J, Lassila O (2001) The Semantic Web. Sci Am 284(5):34–43
22. Bettini C, Brdiczka O, Henricksen K, Indulska J, Nicklas D, Ranganathan A, Riboni D (2010) A survey of context modelling and reasoning techniques. Pervasive Mob Comput 6(2):161–180
23. Biczók G, Fehske A, Malmodin J (2011) Deliverable D2.1 - economic and ecological impact of ICT. EARTH project. https://bscw.ict-earth.eu/pub/bscw.cgi/d38532/EARTH_WP2_D2.1_v2.pdf
24. Bizer C, Cyganiak R, Heath T (2008) How to publish linked data on the web. http://wifo5-03.informatik.uni-mannheim.de/bizer/pub/LinkedDataTutorial/
25. Bizer C, Heath T, Berners-Lee T (2009) Linked data - the story so far. Int J Semant Web Inf Syst 5(3):1–22
26. Bizer C, Lehmann J, Kobilarov G, Auer S, Becker C, Cyganiak R, Hellmann S (2009) DBpedia - a crystallization point for the web of data. Web Semant Sci Serv Agents World Wide Web 7(3):154–165
27. Brickley D (2003) Basic geo (WGS84 lat/long) vocabulary. W3C Semantic Web Interest Group, W3C. https://www.w3.org/2003/01/geo/
28. Brickley D, Miller L (2014) FOAF vocabulary specification 0.99. Namespace document. http://xmlns.com/foaf/spec/
29. Buchholz T, Küpper A, Schiffers M (2003) Quality of context: what it is and why we need it. In: Proceedings of the 10th workshop of the hp openview university association, Geneva, Switzerland, pp 1–14
30. Buil-Aranda C, Hogan A, Umbrich J, Vandenbussche P-Y (2013) SPARQL Web-querying infrastructure: ready for action? In: Proceedings of the 12th international semantic web conference - Part II, ISWC '13, New York, NY, USA. Springer, New York, pp 277–293
31. Cardoso J (2007) The semantic web vision: where are we? IEEE Intell Syst 22(5):84–88
32. Carothers G (2014) RDF 1.1 N-quads. W3C recommendation, W3C. https://www.w3.org/TR/n-quads/
33. Carothers G, Seaborne A (2014) RDF 1.1 trig. W3C recommendation, W3C. https://www.w3.org/TR/trig/
34. Chen G, Kotz D (2000) A survey of context-aware mobile computing research. Technical Report, Hanover, NH, USA
35. Chen H, Finin T, Joshi A (2003) An ontology for context-aware pervasive computing environments. Knowl Eng Rev 18(3):197–207
36. Chen H, Finin T, Joshi A (2005) The SOUPA ontology for pervasive computing. Ontologies for agents: theory and experiences. Birkhäuser Basel, Basel, pp 233–258

37. Chen MY, Sohn T, Chmelev D, Haehnel D, Hightower J, Hughes J, LaMarca A, Potter F, Smith I, Varshavsky A (2006) Practical metropolitan-scale positioning for GSM phones. In: Proceedings of the 8th international conference on ubiquitous computing, UbiComp '06. Springer, Berlin, pp 225–242
38. Cheng Y-C, Chawathe Y, LaMarca A, Krumm J (2005) Accuracy characterization for metropolitan-scale Wi-Fi Localization. In: Proceedings of the 3rd international conference on mobile systems, applications, and services, MobiSys '05. ACM, New York, pp 233–245
39. Chu M (2007) New magical blue circle on your map. http://googlemobile.blogspot.de/2007/11/new-magical-blue-circle-on-your-map.html
40. Cleary D, Danev B, O'Donoghue D (2005) Using ontologies to simplify wireless network configuration. In: Proceedings of the formal ontology meets industry, FOMI '05, Verona, Italy
41. Compton M, Barnaghi P, Bermudez L, García-Castro R, Corcho O, Cox S, Graybeal J, Hauswirth M, Henson C, Herzog A, Huang V, Janowicz K, Kelsey WD, Phuoc DL, Lefort L, Leggieri M, Neuhaus H, Nikolov A, Page K, Passant A, Sheth A, Taylor K (2012) The SSN ontology of the W3C semantic sensor network incubator group. Web Semant Sci Serv Agents World Wide Web 17:25–32
42. Crockford D (2006) The application/json media type for javascript object notation (JSON). RFC 4627, RFC editor. http://www.rfc-editor.org/rfc/rfc4627.txt
43. Cuel R, Delteil A, Louis V, Rizzi C (2007) Knowledge web technology roadmap "The technology roadmap of the semantic web". White Paper. http://knowledgeweb.semanticweb.org/semanticportal/docs/download3305.pdf
44. Davis I (2010) RELATIONSHIP: a vocabulary for describing relationships between people. Namespace document. http://vocab.org/relationship/
45. Deva B, Rodriguez Garzon S, Küpper A (2016) FlashPoll: a context-aware polling ecosystem for mobile participation. In: Proceedings of the 19th international conference innovation in clouds, internet and networks, ICIN '16, IFIP, pp 169–176
46. Dey AK (2001) Understanding and using context. Pers Ubiquitous Comput 5(1):4–7
47. Duerst M, Suignard M (2005) Internationalized Resource Identifiers (IRIs). RFC 3987, RFC editor. http://www.rfc-editor.org/rfc/rfc3987.txt
48. ETSI (2015) Digital cellular telecommunications system (Phase 2+) (GSM); Universal Mobile Telecommunications System (UMTS); LTE; Organization of subscriber data (3GPP TS 23.008 version 13.4.0 Release 13)
49. ETSI (2016). Digital cellular telecommunications system (Phase 2+) (GSM); Universal Mobile Telecommunications System (UMTS); LTE; Network architecture (3GPP TS 23.002 version 13.6.0 Release 13)
50. Faggiani A, Gregori E, Lenzini L, Luconi V, Vecchio A (2014) Smartphone-based crowdsourcing for network monitoring: opportunities, challenges, and a case study. IEEE Commun Magaz 52(1):106–113
51. Fielding RT, Gettys J, Mogul JC, Nielsen HF, Masinter L, Leach PJ, Berners-Lee T (1999) Hypertext Transfer Protocol – HTTP/1.1. RFC 2616, RFC editor. http://www.rfc-editor.org/rfc/rfc2616.txt
52. Floréen P, Przybilski M, Nurmi P, Koolwaaij J, Tarlano A, Luther M, Bataille F, Boussard M, Mrohs B, Lau S (2005) Towards a context management framework for MobiLife. In: Proceedings of the IST mobile and wireless communications
53. Foundation for Intelligent Physical Agents (2002) FIPA device ontology specification. Standard. http://www.fipa.org/specs/fipa00091/
54. Gandon F, Schreiber G (2014) RDF 1.1 XML syntax. W3C recommendation, W3C. http://www.w3.org/TR/rdf-syntax-grammar/
55. Gazzè D, Lo Duca A, Marchetti A, Tesconi M (2015) An overview of the tourpedia linked dataset with a focus on relations discovery among places. In: Proceedings of the 11th international conference on semantic systems, SEMANTICS '15. ACM, New York, pp 157–160
56. Goldberg DW, Wilson JP, Knoblock CA (2007) From text to geographic coordinates: the current state of geocoding. URISA J 19(1):33–47

57. Golemati M, Katifori A, Vassilakis C, Lepouras G, Halatsis C (2007) Creating an ontology for the user profile: method and applications. In: Proceedings of the first international conference on research challenges in information science, RCIS '07, pp 407–412
58. Guha R, Brickley D (2014) RDF schema 1.1. W3C recommendation, W3C. https://www.w3.org/TR/rdf-schema/
59. Halpin H, Herman I, Hayes PJ (2010) When owl:sameAs isn't the same: an analysis of identity links on the semantic web. In: Proceedings of the linked data on the web workshop, LDOW '10
60. Hasan Z, Boostanimehr H, Bhargava VK (2011) Green cellular networks: a survey, some research issues and challenges. IEEE Commun Surv Tutor 13(4):524–540
61. Hausenblas M (2009) Exploiting linked data to build web applications. IEEE Internet Comput 13(4):68–73
62. Heath T, Bizer C (2011) Linked data: evolving the web into a global data space. Synthesis lectures on the semantic web. Morgan & Claypool Publishers, Milton Keynes
63. Heckmann D, Schwartz T, Brandherm B, Schmitz M, von Wilamowitz-Moellendorff M (2005) GUMO – The General User Model Ontology. In: Proceedings of the user modeling 2005: 10th international conference, UM 2005, Edinburgh, Scotland, UK, 24–29 July 2005. Springer, Berlin, pp 428–432
64. Herman I, Adida B, Sporny M, Birbeck M (2015) RDFa 1.1 primer - third edition. W3C note, W3C. http://www.w3.org/TR/rdfa-primer/
65. Hervás R, Bravo J, Fontecha J (2010) A context model based on ontological languages: a proposal for information visualization. J Univers Comput Sci 16(12):1539–1555
66. Hitzler P, Krötzsch M, Rudolph S (2009) Foundations of semantic web technologies, 1st edn. Chapman & Hall/CRC, Boca Raton
67. Hochstatter I, Küpper A, Schiffers M, Köthner L (2003) Context provisioning in cellular networks. In: Proceedings of 8th international workshop on mobile multimedia communications
68. Hossain M (2012) Users' motivation to participate in online crowdsourcing platforms. In: Proceedings of the international conference on innovation management and technology research, ICIMTR '12, pp 310–315
69. Howe J (2008) Crowdsourcing: why the power of the crowd is driving the future of business, 1st edn. Crown Publishing Group, New York
70. Huang C-W, Shih T-Y (1997) On the complexity of point-in-polygon algorithms. Comput Geosci 23(1):109–118
71. Ilarri S, Illarramendi A, Mena E, Sheth A (2011) Semantics in location-based services [Guest editor's introduction]. IEEE Internet Computing, 15(6):10–14
72. International Telecommunication Union (ITU) (2004). List of Mobile Country or Geographical Area Codes (Complement to ITU T Recommendation E.212 (11/98)
73. International Telecommunication Union (ITU) (2014) Mobile Network Codes (MNC) for the international identification plan for public networks and subscriptions (According to Recommendation ITU-T E.212 (05/2008)
74. Jain P, Hitzler P, Janowicz K, Venkatramani C (2013) There's no money in linked data. Technical report, DaSe Lab, Department of Computer Science and Engineering, Wright State University, Dayton, OH, USA
75. Jain P, Hitzler P, Sheth AP, Verma K, Yeh PZ (2010) Ontology alignment for linked open data. In Proceedings of the 9th international semantic web conference on the semantic web - volume part I, ISWC '10, pages 402–417, Berlin, Heidelberg. Springer-Verlag
76. Jones RK, Liu L (2006) What where wi: an analysis of millions of Wi-Fi access points. Technical Report, Georgia Institute of Technology
77. Kim M, Fielding JJ, and Kotz D (2006) Risks of using AP locations discovered through war driving. In Proceedings of the 4th international conference on pervasive computing, PERVASIVE '06, pages 67–82, Berlin, Heidelberg. Springer-Verlag
78. Kim S, Kwon J (2007) Effective context-aware recommendation on the semantic web. Int J Comput Sci Netw Secur 7(8):154–159

79. Knappmeyer M, Kiani SL, Frà C, Moltchanov B, Baker N (2010) ContextML: A Light-Weight context representation and context management schema. In Proceedings of the IEEE 5th international symposium on wireless pervasive computing, ISWPC '10, pages 367–372
80. Knappmeyer M, Kiani SL, Reetz ES, Baker N, Tonjes R (2013) Survey of context provisioning middleware. IEEE Communications Surveys Tutorials, 15(3):1492–1519
81. Kunze CP, Zaplata S, Turjalei M, Lamersdorf W (2008) Enabling context-based cooperation: a generic context model and management system. Business Information Systems. volume 7 of Lecture Notes in Business Information Processing. Springer, Berlin Heidelberg, pp 459–470
82. Kunze SR, Auer S (2013) Dataset retrieval. In Proceedings of the 2013 IEEE seventh international conference on semantic computing, ICSC '13, pages 1–8, Washington, DC, USA. IEEE Computer Society
83. Küpper A (2005) Location-based services: Fundamentals and Operation. John Wiley & Sons, 1st edition
84. Küpper A, Treu G, Linnhoff-Popien C (2006) TraX: a device-centric middleware framework for location-based services. IEEE Commun Magaz 44(9):114–120
85. Lanthaler M, Cyganiak R, Wood D (2014) RDF 1.1 concepts and abstract syntax. W3C recommendation, W3C. http://www.w3.org/TR/rdf11-concepts/
86. Le-Phuoc D, Hauswirth M (2009) Linked open data in sensor data mashups. In: Proceedings of the 2nd international workshop on semantic sensor networks, SSN '09, Aachen, Germany, CEUR-WS.org, pp 1–16
87. Le-Phuoc D, Parreira JX, Hausenblas M, Han Y, Hauswirth M (2010) Live linked open sensor database. In: Proceedings of the 6th international conference on semantic systems, I-SEMANTICS '10. ACM, New York, pp 46:1–46:4
88. Lee K, Lee J, Kwan M-P (2017) Location-based service using ontology-based semantic queries: a study with a focus on indoor activities in a university context. Comput Environ Urban Syst 62:41–52
89. Lehmann L (2012) Location-based mobile games. Seminar paper, Technische Universität Berlin. https://www.snet.tu-berlin.de/fileadmin/fg220/courses/WS1112/snet-project/location-based-mobile-games_lehmann.pdf
90. Lenat D (1998) The dimensions of context-space. Technical Report, Cycorp
91. Li Z, Hongjuan Z (2011) Research of crowdsourcing model based on case study. In: Proceedings of the 8th international conference on service systems and service management, ICSSSM '11, pp 1–5
92. Lieberman J, Singh R, Goad C (2007) W3C geospatial vocabulary. W3C incubator group, W3C. https://www.w3.org/2005/Incubator/geo/XGR-geo/
93. Llanes KR, Casanova MA, Lemus NM (2016) From sensor data streams to linked streaming data: a survey of main approaches. J Inf Data Manag 7(2):130–140
94. Lott R (2015) Geographic information - well-known text representation of coordinate reference systems. OGC Standard 1.0, Open Geospatial Consortium. https://portal.opengeospatial.org/files/12-063r5
95. Makris P, Skoutas DN, Skianis C (2013) A survey on context-aware mobile and wireless networking: on networking and computing environments' integration. IEEE Commun Surv Tutor 15(1):362–386 First Quarter
96. Malhotra A, Thompson H, Biron PV, Peterson D, Gao S, Sperberg-McQueen M (2012) W3C XML Schema Definition language (XSD) 1.1 Part 2: datatypes. W3C recommendation, W3C. https://www.w3.org/TR/xmlschema11-2/
97. Mankowitz JD, Paverd AJ (2011) Mobile device-based cellular network coverage analysis using crowd sourcing. In: EUROCON - international conference on computer as a tool, pp 1–6
98. Mannweiler C, Klein A, Schneider J, Schotten HD (2009) Exploiting user and network context for intelligent radio network access. In: Proceedings of the 2009 international conference on ultra modern telecommunications workshops, ICUMT '09, pp 1–6
99. Manzoor A, Truong H-L, Dustdar S (2008) On the evaluation of quality of context. In: Proceedings of the Smart sensing and context: third european conference, EuroSSC 2008, Zurich, Switzerland, 29–31 October 2008. Springer, Berlin, pp 140–153

100. Manzoor A, Truong H-L, Dustdar S (2009) Using quality of context to resolve conflicts in context-aware systems. Quality of context: first international workshop, QuaCon 2009, Stuttgart, Germany, 25–26 June 2009. Revised papers. Springer, Berlin, pp 144–155
101. Marshall CC, Shipman FM (2003) Which semantic web? In: Proceedings of the fourteenth ACM conference on hypertext and hypermedia, HYPERTEXT '03. ACM, New York, pp 57–66
102. Mascardi V, Locoro A, Rosso P (2010) Automatic ontology matching via upper ontologies: a systematic evaluation. IEEE Trans Knowl Data Eng 22(5):609–623
103. McCullagh D (2011) Microsoft collects locations of Windows phone users. http://www.cnet.com/news/microsoft-collects-locations-of-windows-phone-users/
104. Megiddo N (1982) Linear-time algorithms for linear programming in R3 and related problems. SIAM J Comput 12:759–776
105. Misund G, Holone H, Karlsen J, Tolsby H (2009) Chase and catch - simple as that?: old-fashioned fun of traditional playground games revitalized with location-aware mobile phones. In: Proceedings of the international conference on advances in computer enterntainment technology, ACE '09. ACM, New York, pp 73–80
106. Moltchanov B, Frà C, Valla M, Licciardi CA (2011) Context management framework and context representation for MNO. In: Proceedings of the AAAI workshop on activity context representation: techniques and languages, AAAI-WS '11. AAAI Press, pp 53–58
107. Naboulsi D, Fiore M, Ribot S, Stanica R (2015) Large-scale mobile traffic analysis: a survey. IEEE Commun Surv Tutor 18(1):124–161
108. Neisse R, Wegdam M, van Sinderen M (2008) Trustworthiness and quality of context information. In: Proceedings of the 9th international conference for young computer scientists, ICYCS '08, pp 1925–1931
109. Nurmi P, Bhattacharya S, Kukkonen J (2010) A grid-based algorithm for on-device GSM positioning. In: Proceedings of the 12th ACM international conference on ubiquitous computing, Ubicomp '10. ACM, New York, pp 227–236
110. Ostuni V, Gentile G, Noia T, Mirizzi R, Romito D, Sciascio E (2013) Mobile movie recommendations with linked data. In: Availability. reliability, and security in information systems and HCI. Lecture notes in computer science, vol 8127. Springer, Berlin, pp 400–415
111. Parent C, Spaccapietra S, Renso C, Andrienko G, Andrienko N, Bogorny V, Damiani ML, Gkoulalas-Divanis A, Macedo J, Pelekis N, Theodoridis Y, Yan Z (2013) Semantic trajectories modeling and analysis. ACM Comput Surv 45(4):42:1–42:32
112. Park TH, Kwon O (2007) Identifying a generic model of context for context-aware multi-services. In: Proceedings of the ubiquitous intelligence and computing: 4th international conference, UIC 2007, Hong Kong, China, 11–13 July 2007. Springer, Berlin, pp 919–928
113. Patkos T, Bikakis A, Antoniou G, Papadopouli M, Plexousakis D (2007) A semantics-based framework for context-aware services: lessons learned and challenges. In: Proceedings of the 4th international conference on ubiquitous intelligence and computing, UIC '07. Springer, Berlin, pp 839–848
114. Patni H, Henson C, Sheth A (2010) Linked sensor data. In: International symposium on collaborative technologies and systems, CTS '10, pp 362–370
115. Perera C, Zaslavsky A, Christen P, Georgakopoulos D (2014) Context aware computing for the internet of things: a survey. IEEE Commun Surv Tutor 16(1):414–454. First Quarter
116. Perry M, Herring J (2012) OGC GeoSPARQL - a geographic query language for RDF data. OGC Standard 1.0, Open Geospatial Consortium. https://portal.opengeospatial.org/files/?artifact_id=47664
117. Polo L, Mínguez I, Berrueta D, Ruiz C, Gómez-Pérez JM (2014) User preferences in the web of data. Semant Web J 5(1):67–75
118. Portele C (2007) OpenGIS Geography Markup Language (GML) encoding standard. OpenGIS Standard 3.2.1, Open Geospatial Consortium. http://portal.opengeospatial.org/files/?artifact_id=20509
119. Poveda Villalon M, Suárez-Figueroa MC, García-Castro R, Gómez-Pérez A (2010) A context ontology for mobile environments. In: Proceedings of the second workshop on context, information and ontologies, CIAO '10, vol 626. CEUR-WS.org

120. Qiao X, Li X, Fensel A, Su F (2011) Applying semantics to Parlay-based services for telecommunication and internet networks. Cent Euro J Comput Sci 1(4):406–429
121. Raimond Y, Schreiber G (2014) RDF 1.1 primer. W3C working group note, W3C. https://www.w3.org/TR/rdf11-primer/
122. Rayfield J (2012) Sports refresh: dynamic semantic publishing. BBC Internet Blog. http://www.bbc.co.uk/blogs/bbcinternet/2012/04/sports_dynamic_semantic.html
123. Rayfield J (2014) Semantic technology for online, broadcast and print media. In: Proceedings of the 4th international conference on web intelligence, mining and semantics, WIMS '14. ACM, New York, pp 3:1–3:2
124. Riboni D, Bettini C (2011) COSAR: hybrid reasoning for context-aware activity recognition. Pers Ubiquitous Comput 15(3):271–289
125. Rodríguez J, Bravo M, Guzmán R (2013) Multidimensional ontology model to support context-aware systems. In: Proceedings of the AAAI workshop on activity context-aware system architectures, AAAI-WS '13. AAAI Press, pp 53–60
126. Rodriguez Garzon S, Deva B (2014) Geofencing 2.0: taking location-based notifications to the next level. In: Proceedings of the 2014 ACM international joint conference on pervasive and ubiquitous computing, UbiComp '14. ACM, New York, pp 921–932
127. Royce WW (1987) Managing the development of large software systems: concepts and techniques. In: Proceedings of the 9th international conference on software engineering, ICSE '87, Los Alamitos, CA, USA. IEEE Computer Society Press, pp 328–338
128. Rula J, Bustamante FE (2012) Crowd (Soft) control: moving beyond the opportunistic. In: Proceedings of the twelfth workshop on mobile computing systems and applications, HotMobile '12. ACM, New York, pp 3:1–3:6
129. Salas JM, Harth A (2011) NeoGeo vocabulary: defining a shared RDF representation for GeoData. Public Draft, NeoGeo. http://geovocab.org/doc/survey.html
130. Sathe S, Melamed R, Bak P, Kalyanaraman S (2014) Enabling location-based services 2.0: challenges and opportunities. In: Proceedings of the IEEE 15th international conference on mobile data management, MDM '14, vol 1. IEEE, pp 317–320
131. Sauermann L, Cyganiak R (2008) Cool URIs for the semantic web. W3C interest group note, W3C. https://www.w3.org/TR/cooluris/
132. Sauter M (2011) From GSM to LTE: an introduction to mobile networks and mobile broadband, 1st edn. Wiley, Oxford
133. Scerri S, Attard J, Rivera I, Valla M, Handschuh S (2012) DCON: Interoperable Context Representation for Pervasive Environments. In: Proceedings of the AAAI workshop on activity context representation: techniques and languages, AAAI-WS '12. AAAI Press, pp 90–97
134. Schilit B, Adams N, Want R (1994) Context-aware computing applications. In: Proceedings of the 1994 First workshop on mobile computing systems and applications, WMCSA '94, Washington, DC, USA. IEEE Computer Society, pp 85–90
135. Schmidt A, Beigl M, Gellersen H-W (1998) There is more to context than location. Comput Graph 23:893–901
136. Schneider J, Klein A, Mannweiler C, Schotten HD (2010) Environmental context detection for context-aware systems. ICaST - ICST's Global Community Magazine. https://icastdev.wordpress.com/2010/02/05/environmental-context-detection-for-context-aware-systems/
137. Schreiber G, Dean M (2004) OWL Web Ontology Language reference. W3C recommendation, W3C. https://www.w3.org/TR/owl-ref/
138. Seaborne A, Carothers G (2014) RDF 1.1 N-triples. W3C recommendation, W3C. https://www.w3.org/TR/n-triples/
139. Seaborne A, Harris S (2013) SPARQL 1.1 query language. W3C recommendation, W3C. https://www.w3.org/TR/sparql11-query/
140. Setten M, Pokraev S, Koolwaaij J (2004) Context-aware recommendations in the mobile tourist application COMPASS. In: Adaptive hypermedia and adaptive web-based systems. Lecture notes in computer science, vol 3137. Springer, Berlin, pp 235–244
141. Sewell B (2010) Apple inc.'s response to request for information regarding its privacy policy and location-based services. http://www.wired.com/images_blogs/gadgetlab/2011/04/applemarkeybarton7-12-10.pdf

142. Skyum S (1991) A simple algorithm for computing the smallest enclosing circle. Inf Process Lett 37:121–125
143. Sporny M, Kellogg G, Lanthaler M (2014) JSON-LD 1.0. W3C recommendation, W3C. https://www.w3.org/TR/json-ld/
144. Stadler C, Lehmann J, Höffner K, Auer S (2012) LinkedGeoData: a core for a web of spatial open data. Semant Web J 3(4):333–354
145. Stenovec T (2015) Google has gotten incredibly good at predicting traffic - here's how. Business Insider. http://www.businessinsider.com/how-google-maps-knows-about-traffic-2015-11
146. Strang T, Linnhoff-Popien C (2004) A context modeling survey. In: UbiComp 1st international workshop on advanced context modelling, reasoning and management, Nottingham, pp 31–41
147. Taylor K, Cox S, Janowicz K, Phuoc DL, Haller A, Lefrançois M (2017) Semantic sensor network ontology. Candidate recommendation, W3C. https://www.w3.org/TR/vocab-ssn/
148. Toutain F, Ramparany F, Szczekocka E (2012) Semantic context reasoning for formulating user location. In: Proceedings of the 26th international conference on advanced information networking and applications workshops, WAINA '12, pp 671–677
149. Tummarello G, Delbru R, Oren E (2007) Sindice.com: weaving the open linked data. In: The semantic web. Lecture notes in computer science, vol 4825. Springer, Berlin, pp 552–565
150. Villalonga C, Strohbach M, Snoeck N, Sutterer M, Belaunde M, Kovacs E, Zhdanova A, Goix LW, Droegehorn O (2009) Mobile ontology: towards a standardized semantic model for the mobile domain. Service-oriented computing - ICSOC 2007 workshops. Springer, Berlin, pp 248–257
151. Volz J, Bizer C, Gaedke M, Kobilarov G (2009) Discovering and maintaining links on the web of data. In: Proceedings of the 8th international semantic web conference, ISWC '09. Springer, Berlin, pp 650–665
152. W3C OWL Working Group (2012) OWL 2 Web Ontology Language document overview (second edition). W3C recommendation, W3C. http://www.w3.org/TR/owl-overview
153. Walsh N, Jacobs I (2004) Architecture of the World Wide Web, vol 1. W3C recommendation, institution = W3C. https://www.w3.org/TR/webarch/
154. Wang XH, Zhang DQ, Gu T, Pung HK (2004) Ontology based context modeling and reasoning using OWL. In: Proceedings of the second IEEE annual conference on pervasive computing and communications workshops, PERCOMW '04, pp 18–22
155. Wick M (2011) GeoNames. In: Symposium on space-time integration in geography and GIScience. https://cga-download.hmdc.harvard.edu/publish_web/2011_AAG_Gazetteer/Wick.ppt
156. Xiuquan Qiao XL, Chen J (2012) Telecommunications service domain ontology: semantic interoperation foundation of intelligent integrated services. In: Telecommunications networks - current status and future trends. InTech, pp 183–210
157. Yang D-H, Bilaver L, Hayes O, Goerge R (2004) Improving geocoding practices: evaluation of geocoding tools. J Med Syst 28(4):361–370
158. Yu HQ, Zhao X, Reiff-Marganiec S, Domingue J (2012) Linked context: a linked data approach to personalised service provisioning. In: Proceedings of the 2012 IEEE 19th international conference on web services, pp 376–383
159. Yu L, Liu Y (2013) Using linked data in a heterogeneous sensor web: challenges, experiments and lessons learned. Int J Digit Earth 8(1):17–37
160. Yürür O, Liu CH, Sheng Z, Leung VCM, Moreno W, Leung KK (2016) Context-awareness for mobile sensing: a survey and future directions. IEEE Commun Surv Tutor 18(1):68–93 First Quarter
161. Zhang WS, Xu N, Yang HD, Zhang XG, Xing X (2013) CACOnt: a ontology-based model for context modeling and reasoning. In: Instruments, measurement, electronics and information engineering. Applied mechanics and materials, vol 347. Trans Tech Publications, 10, pp 2304–2310
162. Zhao J, Alexander K, Cyganiak R, Hausenblas M (2011) Describing linked datasets with the VoID vocabulary. W3C note, W3C. http://www.w3.org/TR/void/

163. Zhou P, Zheng Y, Li Z, Li M, Shen G (2012) IODetector: A generic service for indoor outdoor detection. In: Proceedings of the 10th ACM conference on embedded network sensor systems, SenSys '12. ACM, New York, pp 113–126
164. Zimmermann A, Lorenz A, Oppermann R (2007) An operational definition of context. In: Proceedings of the 6th international and interdisciplinary conference on modeling and using context, CONTEXT '07. Springer, Berlin, pp 558–571

Printed in the United States
By Bookmasters